Ethnoecology

Ethnoecology

Situated Knowledge/Located Lives

EDITED BY

Virginia D. Nazarea

THE UNIVERSITY OF ARIZONA PRESS TUCSON

The University of Arizona Press

www.uapress.arizona.edu

First paperbound printing 2003

Library of Congress Cataloging-in-Publication Data
Ethnoecology : situated knowledge / located lives / edited by
Virginia D. Nazarea.
p. cm.
Includes index.
ISBN 978-0-8165-2364-1 (pbk : acid-free paper)
1. Ethnobiology. 2. Human ecology. 3. Traditional farming.
4. Biotic communities. 5. Indians—Ethnobiology.
I. Nazarea, Virginia D. (Virginia Dimasuay), 1954–
GN476.7.E77 1999 98-25481
304.2—ddc21

Manufactured in the United States of America on acid-free, archival-quality paper.

15 14 13 12 11 10 8 7 6 5 4 3

Contents

Preface

Ethnoecology is "a way of looking" at the relationship between humans and the natural world that emphasizes the role of cognition in framing behavior. Without doubt, it is a powerful perspective from which to understand resource recognition and management: the schemas, scripts, and action plans that orient people in the world and determine the productivity, equity, and sustainability of their practices. Hence, conceptually and methodologically, ethnoecology offers great promise for linking anthropology meaningfully with other fields of investigation and discourse.

Somewhere along the line, however, "a way of looking" was transformed to "*the* way of looking," thus imbuing cultural, disciplinary, and, to some extent, ideological filters with an omniscience, universality, and immutability that impart some comfort but at the same time paralyze, empower but at the same time caricature and parody. The reduction of dynamic, situated systems of understanding embedded in history, politics, and environment to homogenous, pan-human categories and systems of classification bereft of any meaningful context has distracted us from the implications and repercussions of different latitudes and points of view on the local capacity for self-determination, on the nurturance of diversity, and even more fundamental, on creative choices balancing persistence and change.

This multidisciplinary volume is a call for recognizing the different takes that social and natural scientists, as well as activists and development practitioners, have on the human dimension of the conservation of natural resources. More than this, it explores the differing viewpoints that local people themselves have on their environment, as shaped by their particular vantage points. Those vantage points, the volume demonstrates, are no accident; they are systematic, both enabling and constraining, and are pivotal to the appreciation of decision-making processes and behavioral outcomes. Ethnoecology offers a useful framework for analyzing problems related to environmental management, agricultural sustainability, biodiversity conservation, and intellectual property rights, as well as furthering a basic understanding of humans-in-environment. But in order for ethnoecology to realize its full potential, we must move toward a more dynamic and context-sensitive approach—an approach that would enable us to admit different takes.

We have been distracted long enough, our energies diverted toward deciphering or imagining regularities and uniformities when in fact the irregularities and variations are infinitely more powerful and relevant. We have been forcing our ideas and those of our informants and collaborators into paradigmatic monocultures—into some form of hypercoherence—when in fact it is the variability, in our concepts as well as in theirs, that constitutes the challenge. Sad to say, we have also been talking mostly with ourselves, and thus have missed out on opportunities to test our abstractions against exciting developments in other fields and real world situations that would help us to develop a tolerance for "mezzo" theories that are not quite so encompassing or so elegant.

The contributors to this volume took time off from their busy schedules in different parts of the globe to participate in the conference entitled "Ethnoecology: Different Takes and Emergent Properties," organized and hosted by the Ethnoecology/Biodiversity Laboratory on April 7 and 8, 1995. If they found the topic fuzzy, if they found the format unsettling, they did not let on. My deep appreciation goes to all of them for being sensitive to the significance of this undertaking and lending their knowledge and experience to the conference and this collection. The Office of the Vice Chancellor for Academic Affairs at the University of Georgia funded the conference through a State-of-the-Art grant, and I would like to thank Dr. Robert Rhoades for his endorsement and Dr. William Prokasy for his support. The State Botanical Garden of Georgia, the USDA Plant Genetic Resources Research Institute, the Sustainable Agriculture and Natural Resources Collaborative Research Support Program, the Institute of Ecology, and the Anthropology Student Organization all co-sponsored the conference and were active partners every step of the way.

I would also like to acknowledge the University of Georgia, especially the Office of the Vice President for Academic Affairs and Provost (Dr. Karen Holbrook), Office of the Dean of the Franklin College (Dr. Wyatt Anderson and Dr. Hugh Ruppersburg), and the Sustainable Agriculture and Natural Resources Management Collaborative Research Support Program's Comparative Ethnoecology Project (Dr. Robert Rhoades), for providing moral and financial support through write-up and publication of this volume.

Drs. Jack Harlan, James Affolter, William Hargrove, and David Newman shared their valuable insights in the various sessions. They viewed ethnoecology from a different disciplinary lens—providing the perspectives of a geneticist, a taxonomist, a soil scientist, and a resource economist, respectively—and enriched the ensuing discussions. For the most part, though, as was the case for the contributors, their most prominent shared attribute was their long-term engagement with various facets of ethnoecology in their own work. Finally, I thank my students and colleagues at the Department of Anthropology who have kept me, always, on my toes and whose collaboration made this book possible. In particular, Joanne Goodwin and Eleanor Tison infused the preparation for and conduct of the conference with energy, dedication, and fresh vision. My special thanks go to Eleanor's twins for cooperating as long as they did and to LaBau Bryan for tying it all up when the twins decided it was time.

This volume is written for the new cohort of ethnoecologists—one student who comes to mind liked ecology and world systems theory, as well as homegardens, deconstruction, and ethnobotany (and managed not to get too confused)—that they may rest solidly on their ancestors' shoulders but not be afraid to venture forth and muddle through. To acknowledge a debt of gratitude for his pioneering work and generous spirit, this work is dedicated to Dr. Harold Conklin, who is many things to many people—pioneering thinker, meticulous fieldworker, unrelenting investigator, indefatigable letter-writer, teacher, friend—on the occasion of his formal retirement from academe.

Abbreviations

ANILCA	Alaska National Interest Lands Conservation Act
ARS	Agricultural Research System
ASCS	Agricultural Stabilization and Conservation Service
BATH	Belize Association of Traditional Healers
BMG	Battle Mountain Gold
CAA	Código Alimentario Argentino
CBD	Convention on Biological Diversity
CCCD	Costilla County Conservancy District
CGIAR	Consultative Group for International Agricultural Research
CHD	coronary heart disease
CIKARD	Center for Indigenous Knowledge for Agriculture and Rural Development
COICA	Coordinating Body for the Indigenous Peoples' Organization of the Amazon Basin
CRP	Conservation Reserve Program
DNA	deoxyribonucleic acid
EOS	Earth Observing System
FAO	Food and Agriculture Organization
FNA	Farmacopea Nacional Argentina
FV	folk crop variety
GEMS	Global Environmental Monitoring System
GPS	global positioning system
GRID	Global Resource Information Database
HRM	Holistic Resource Management
HUD	Department of Housing and Urban Development
HYV	high yielding variety
IAITP	International Alliance of the Indigenous-Tribal Peoples of the Tropical Forests
IBPGR	International Board for Plant Genetic Resources

ICBP	International Cooperative Biodiversity Program
IK	indigenous knowledge
ILO	International Labor Organization
IPGRI	International Plant Genetic Resources Institute
IPR	intellectual property rights
IRRI	International Rice Research Institute
ISE	International Society of Ethnobiology
MLRA	Mined Land Reclamation Act
NEH	National Endowment for the Humanities
NGO	nongovernmental organization
NIDDM	non-insulin-dependent diabetes mellitus
NPGS	National Plant Germplasm System
NRC	National Research Council
NRCS	Natural Resources and Conservation Service
PCB	polychlorinated biphenyl
PRODEMA	Programa para el Desarrollo de Plantas Medicinales y Aromáticas
PTC	phenylthiocarbamide
RAFI	Rural Advancement Foundation International
SAIIC	South and Mesoamerican Indian Information Center
SCS	Soil Conservation Service
TAT	Thematic Apperception Test
TEK	traditional environmental knowledge
TRR	traditional resource rights
TWN	Third World Network
UN	United Nations
UNDP	United Nations Development Programme
UNESCO	United Nations Educational, Scientific and Cultural Organization
USDA	U.S. Department of Agriculture
USFS	U.S. Forest Service
WCIP	World Council of Indigenous Peoples
WIPO	World Intellectual Property Organization

Ethnoecology

1

Introduction

A View from a Point: Ethnoecology as Situated Knowledge

VIRGINIA D. NAZAREA

In 1954, Harold Conklin wrote his dissertation on "The Relation of the Hanunuo Culture to the Plant World." In the same year, he introduced what he called "the ethnoecological approach" in a seminal paper that was to dismantle the dominant view on shifting cultivation as a haphazard, destructive, and primitive way of making a living. What came after, from the midfifties to the midseventies, was a testimony to the power of the idea that Conklin had unleashed (for useful reviews, see Hunn 1989; Ford 1978; Fowler 1977; Toledo 1992). The prefix "ethno" came to denote not merely a localized application of a branch of study (for example, ethnobotany as the botany of a local group from an outsider's—that is, an investigator's—perspective) but also, following the works of Conklin (1954, 1961), Goodenough (1957), Frake (1962), Sturtevant (1964), and many others, a serious attempt toward the understanding of local understanding(the so-called native point of view)about a realm of experience. An explosion of research papers, not to mention entire programs at prestigious universities, systematically documented and analyzed folk classification and paradigms pertaining to plants, animals, firewood, soils, water, illness, and the human body until only the most incorrigible could remain unimpressed by the logic, complexity, and sophistication of local knowledge.

Anthropologists and nonanthropologists alike could not stop marveling at why, to use Brent Berlin's phrase (1992:5), "non-literates 'know so much' about nature." This sense of amazement and perplexity has been pursued, broadly speaking, in two different directions. One, as exemplified by Conklin's original conception of ethnoecology, is to demonstrate Western scientific ignorance about other peoples' ways of thinking and doing, and to point out its arrogance in dismissing anything that is different as being inferior. The other, as exemplified by the methodical investigation of Tzeltal ethnobotany by Berlin, Breedlove, and Raven (1974) is to cross-refer native systems of classification to the Western scientific tradition—in this case, the Linnaean taxonomic system—and to demonstrate how native systems virtually match scientific taxonomies rank by rank, category by category.

Both approaches led to a qualitative leap in the way local knowledge is regarded, causing a radical shift in mindset from viewing native systems of thought as naive and rudimentary, even savage, to a recognition that local cultures know their plant, animal, and physical resources intimately and are expert at juggling their options for meeting day-to-day requirements and making the most of ephemeral opportunities. Ethnoscience introduced a methodological rigor and theoretical depth that had been quite unknown in past cataloguing of the local uses of biological resources. There is a difference between the two approaches, however, if not by intent then at least by implication. I would argue that the first approach places value on local knowledge by reference to its internal coherence and its environmental and sociocultural adaptiveness. In contrast, the second approach strives to demonstrate the primacy of perceptual universals in determining patterns of classification. In so doing, it subjects local knowledge to a test of legitimacy by measuring it against Western systems of classification and downplaying its adaptability to varying environmental demands and cultural dimensions that have shaped, and continue to shape, its many formulations.

The distinction between these two trajectories is not petty, and the problem needs to be discussed because of contemporary concerns about the representation of local knowledge and related issues of authorship, access, and control. These issues inform, or should inform, national, regional, and international negotiations about biodiversity and the commons and about self-determination and intellectual property rights, as well as our understanding of humans-in-environment. Gone are the simpler days when anthropologists could refer to their fieldwork sites as "my village" and speak authoritatively about "my people," or use Western systems of thought as the yardstick for everything that is good and beautiful and true. As Gary Lease (1995:5) perceptively noted:

> In our post-modern, post-Marxist world, class struggles no longer have anything to do with "truth," with "right" and "wrong," but

> rather only with the most profound level of ideological battles. . . . Such contests never result in victory, in completion, in closure. We will not "get the story right," regardless of the tendency of some scientists to proclaim final triumph. . . . Our many representations of nature and human are, in other words, always and ultimately flawed. . . . This, in turn, underlines the role of *power* in the contestation over what gets to count in any ruling narrative, and who gets to tell it.

There is another, related level in which the debate has been pursued, this time more openly. This concerns the question about whether systems of classification are intellectually driven, a natural pan-human response to being confronted by the chaos (Lévi-Strauss 1966) or the chunks of biological diversity (Berlin, Breedlove, and Raven 1974), or motivated primarily by the utilitarian concerns of human beings as biological entities themselves who need to eat, sleep, keep warm, seek shelter, defend their plots, heal, and reproduce (Hunn 1982). Berlin made his position clear:

> One is not able to look out on the landscape of organic beings and organize them into cultural categories that are, at base, inconsistent with biological reality. The world of nature cannot be viewed as a continuum from which pieces may be selected ad libitum and organized into arbitrary cultural categories. Rather, groups of plants and animals present themselves to the human observer as a series of discontinuities whose structure and content are seen by all human beings in essentially the same ways, perceptual givens that are largely immune from the variable cultural determinants found in other areas of human experience. (1992:8–9)

As a counterpoint, Hunn's observation about the striking difference between the minimal classificatory effort directed by the Tzeltal to adult butterflies that do not significantly affect their livelihood, and the considerable attention—resulting in more complicated classificatory schemes—they devote to caterpillars that do, indicates that in fact other areas of human experience impact classification in quite significant and interesting ways (Hunn 1982).

Distinct, but in close affinity to the second position, is the emphasis on cultural relations that shape classifications—an argument espoused, for example, by Ellen (1993)—that also questions the disembodied universalist, intellectualist stance. In explaining his position, Ellen wrote:

> My own intellectual socialization within the British tradition of social anthropology had brought with it an empirical and sociological bias which militated against an approach which seemed to me to reduce "mundane" classifications to narrow intellectual conundrums to be solved through the application of formal mathematical, logical, and linguistic procedures, or which relegated their analysis to comparative and evolutionary speculation about general mental principles of classification or cognition. . . . Without denying the importance of these matters, my main theoretical concern has been with classifications as situationally adapted and dynamic devices of practical importance to their users, reflecting an interaction—though in a by no means self-evident way—between culture, psychology, and discontinuities in the concrete world; a lexical and semantic field firmly embedded in a wider context of beliefs and social practices. (1993:3)

My purpose in organizing the conference entitled "Ethnoecology: Different Takes and Emergent Properties," which led to this volume, was not to add yet another dissenting voice to this venerable debate. To my mind, the main protagonists in this debate are trying to answer different questions, and, although much has been accomplished in extolling local knowledge and paying respect to its authors, an inordinate amount of energy has already been devoted to arguing for the best possible answer—to "get the story right" once and for all—to sets of questions that are fundamentally different to start with. Berlin has focused his efforts on elucidating universals based on his premise that ethnobiological classification is perceptually driven, while Hunn, Ellen, and others are more concerned with how culture shapes cognition and mediates behavior. There is no reason why human beings cannot operate at both levels sequentially or even simultaneously, as, I think, perhaps they do. In the meantime, we may be missing the opportunity to move on and pursue other interesting directions, to connect intellectually with exciting dialogues within and outside anthropology, and to address real world concerns that are larger than our limited, albeit intense, paradigms.

I believe it is time to reorient the conversation to focus on an important dimension that has largely been missed, a problem with which ethnoecology has great potential for productive engagement, both at the theoretical and at the applied level. I refer to the connection between plant classification, for example, and conservation of plant genetic resources, or between cultural conceptions of landscape and management of the commons. In short, it is time to turn our attention to the interface between cognition and action—or decision making frameworks and behavioral outcomes—and the lenses and latitudes that shape and structure these interconnections. We can only begin to tackle this problem, however, if we shift our attention from relations of similarity or paradigmatic alliances captured by our neat but static taxonomic trees to relations of contiguity embracing both syntagmatic and diachronic flow.

In an earlier conceptual paper, Hunn (1989:147) referred to this distinction as the Image vis-á-vis the Plan and noted that while "cognitive anthropologists have made substantial progress in the analysis of cultural Image, of Image domains such as color, kinship relations, folk biological taxonomies, and folk anatomy . . . what is lacking is an effective integration of our models of Image and of Plan." Such integration would enable us to link categories to strategies and decipher the "action plans" and "activity signatures" (Randall and Hunn 1984) embedded in each category—a crucial step in understanding the role of local knowledge in human-environment interaction. We may also recall that while Conklin applied linguistic analysis to the service of describing spheres of local knowledge or semantic domains, he never lost sight of linkages between cognition, decision making, and action, or the embeddedness of ethnoecological systems in the environmental and cultural matrix. Discussing the importance of the "cultural axis," for instance, Conklin emphasized that:

> Along the cultural axis, three distinctions are noted: technological, social, and ethnoecological. Technological factors refer to the ways in which the environment is artificially modified, including the treatment of crops, soils, pests, etc. In systems of shifting cultivation, these relationships are of primary importance and often exhibit great complexity; . . . Social factors involve the sociopolitical organization of the farming population in terms of residential, kin, and economic groups. These factors are usually well within the domain of anthropological interest. Ethnoecological factors refer to the ways in which environmental components and their interrelations are categorized and interpreted locally. Failure to cope with this aspect of cultural ecology, to distinguish clearly between native environmental categories (and associated beliefs) and those used by the ethnologist, can lead to confusion, misinformation, and the repetition of useless cliches in discussing unfamiliar systems of land use. (1961:6)

Incorporating contiguity and process as critical components of an engaged ethnoecology also moves us closer to a dynamic rather than monolithic ethnoecology that will admit the importance of ideological negotiation and positioning. No longer encumbered by the need to essentialize our native collaborators, or freeze their taxonomies—or artifacts thereof—in time and space, we can better appreciate how understanding is shaped by standing, as is disposition by position, in an internally differentiated hierarchy of social, economic, and political relations. We can weave into our analysis the history of asymmetric relations with reference to class, gender, and ethnicity, a history that is all too easy to forget if we confine our analysis to perceptual givens, but a history that cannot be finessed because it continues to shape the present. Current thinking in psychology supports the position that even perception is "intelligent"—that it is based on a mental template that incorporates experience and socialization and makes the interpretation of what is perceived a nonmechanical, nonrandom process (Banks and Krajicek 1991). Since it is impossible to maintain that the formation of our mental templates occurs in a social vacuum, the "programming," in a qualified sense, of perception by constraints imposed by our social niche makes rods-and-cones determinism untenable. D. W. Meinig, a noted geographer, actually preceded the psychologists in articulating this insight:

> It will soon be apparent that even though we gather together and look in the same direction at the same instant, we will not—we cannot—see the same landscape. We may certainly agree that we will see many of the same elements—houses, roads, trees, hills—in such denominations as number, form, dimension, and color but such facts take on meaning only through association; they must be fitted together according to some coherent body of ideas. Thus we confront the central problem: any landscape is comprised not only of what lies before our eyes but what lies inside our heads. (1979:33)

The papers in this volume encompass a distillation of research and thinking by many individuals in ethnoecology and related disciplines who address such questions as these: How are folk (and scientific) models shaped, and for what ends? Who defines niches for different groups? Why do cognitive maps vary? By what processes and means is knowledge "naturalized"? In other words, following Meinig, how does "what lies inside our heads" structure how we see and act upon "what lies before our eyes"? Ethnoecology, as the investigation of systems of perception, cognition, and the use of the natural environment, can no longer ignore the historical and political underpinnings of the representational and directive aspects of culture, nor turn away from issues of distribution, access, and power that shape knowledge systems and the resulting practices. In searching for answers and directions, we are guided by Bourdieu's admonition that the social scientist cannot operate under the illusion that he or she can ever hope to produce "an account of accounts," since: "In reality, agents are both classified and classifiers. But they classify according to (or depending upon) their position within classifications. To sum up what I mean by this, I can comment briefly on the notion of point of view: the point of view is a perspective, a partial, subjective vision. . . . But it is at the same time a view, a perspective, taken from a point, from a determinate position in an objective social space (1987:2)."

How are human actors, long regarded by philosophers and ethnoscientists as driven classifiers, themselves classified? Pegged, or more precisely, primed as either leader or led, masculine or feminine, authority or apprentice, how do human agents fabricate their lenses and retool those of others? By what mechanisms are positions defined and secured or negotiated and challenged? From the presentation of papers to the discussions that ensued, it was apparent that the contributors were concerned with local knowledge as situated and distributed, the axes and causes of such asymmetry, and the implications of locality, memory, and sense of place, both for the knowledge bearer and to those who would seek to secure access to such knowledge in the name of the common good, partisan ends, or individual gain. Current concerns with sustainability and the use of biological resources for human welfare and conflicting interests in development and conservation at various levels make it increasingly difficult to ignore these questions. Growing recognition of the cultural, one might even say ideological, construction of natural resources and the appropriation of the commons makes it more imperative for us to re-examine our models—both anthropological and ecological. For all the diversity in disciplinary orientation and epistemological leanings represented in the conference, the contributors shared an appreciation for the value of local environmental knowledge in and of itself, the highly politicized reality in which this knowledge is sustained and manipulated, and the larger goal of cultural and biological persistence in a fragile and contested global commons.

I have organized the present volume into four parts that reflect clusters of concerns expressed by the contributing authors. Part I emphasizes the connection of past to present, the importance of cultural memory and sense of place as opposed to the "politics of disappearance," and the peril that faces those who choose to ignore this most basic of ecological principles: "Everything is connected to everything else." Part II focuses on the existence of multiple viewpoints that refract reality at different angles, thereby causing significant displacements from our normative assumptions. These displacements are of two sorts. As the papers will show, there is a parallax at the level of the operational but this also results in, or from, a parallax at the level of the conceptual.

In Part III, the authors try to untangle issues of scale, tradeoffs, and disenfranchisement as they examine problems related to the commons and ethnoecology's role in analyzing and helping to resolve some of these dilemmas. As neoclassical approaches become more and more futile in dealing with commons issues, where do we search for guidance in evaluating the traditional uses of commons resources versus the market potential of such resources? How do we gauge the sustainability and equitability of scaling up extraction or production from the commons? Without presuming to hold all the answers, ethnoecology's relevance in the context of the present is highlighted in Part IV. This final section pulls together three related dimensions of "the commons"—land and associated resource clusters, landraces or folk varieties, and local or indigenous knowledge—and assesses ongoing efforts at conserving the totality of this priceless human heritage, as well as protecting the vulnerable wellsprings of this legacy.

Ethnoecology: Cultural Memory and Sense of Place

Traditional environmental knowledge, or TEK, according to Eugene Hunn, consists of independent alternatives to the global market syndrome. However, he notes a certain degree of irony in "exploiting TEK to support a global system that is deeply implicated in its destruction." Tackling yet another sensitive issue, Hunn argues for an in vivo rather than in vitro preservation of TEK by supporting the persistence of subsistence strategies among communities that exhibit a high degree of "rootedness" in the local environment. He cites the example of the Alaskan National Interest Land Conservation Act, which allowed for continuing subsistence harvests in Alaskan national parks by adjacent rural communities. This has caused a major cognitive dissonance on the part of park administrators and staff who associate park status with complete absence of human impact (with the exception of tourists). A substantial research program akin to applied ethnoecology was mounted to understand the nature and meaning of subsistence for these communities. One conclusion it came up with was that "the meaning of subsistence is different for each community, varying with ethnic, religious, and economic histories of communities and their component families."

Going right to the heart of the matter, Lillie Lane's evocative essay "The Practical and Religious Meanings of the Navajo Hogan" contrasts standard outsider perspectives with an insider's viewpoint on the traditional Navajo dwelling. According to Lane, "To the outsider, hogans may seem crude structures because they lack modern conveniences such as running water, electricity, and plumbing. Nevertheless, the Navajo hogan is a structure that embodies the social, religious, and philosophical teachings of the Dine' " (as the Navajo refer to themselves). The cultural significance of the hogan lies in its power to locate individuals within the universe and in relation to sacred mountains, shepherding areas, and watering holes. The orientation of hogans also embodies pivotal relationships and symbolisms such as Mother Earth in relation to Father Sun, reflecting, at the same time, female responsibilities (reproduction) vis-à-vis male responsibilities (production). The hogan is therefore, in a very concrete sense, a locating device for the Navajo sense of self and a vital thread in the persistence of cultural memory.

Pursuing a related theme, Michael Dove alludes to the tendency of development programs to banish the old to make way for the new in mindless pursuit of modernization. Asserting that "the 'disappearing' approach of the Western developmental paradigm leaves a false image of agriculture without a history, and thus, without alternatives," Dove describes the case of the Dayaks in Indonesia who, through the ritual maintenance of archaic plants such as Job's tears and millet in their swidden fields, sustain the "presence of the past in the present." Dove's paper presents original, highly provocative insights into two challenges we face today. One is the cognitive challenge of transcendence of present constraints toward greater self-determination through the preservation of local options. This resistance to hegemony through the persistence of memory should be the focus of an engaged ethnoecology, or an emerging political ecology of cognition. The other is the need to re-examine research focused solely on the material and mechanical aspects of sustainability for possible neglect of ideological factors which, as the persistence of archaic plants demonstrates, may in fact play a more powerful role.

The linkage between the ideological and the material is also evident in Zuni land management. Richard Ford uses ethnoscientific methods to elaborate on the extensive knowledge base of the Zuni with reference to soils, water, and plants, and he explains how this knowledge informs Zuni resource management and conservation practices. From this perspective, he demonstrates why a government irrigation project that did not take into account the Zuni's intricate knowledge of agronomy and water management was bound to fail. In the process of summoning evidence to support Zuni claims regarding the thoughtless destruction of their land, Ford observed that Zuni, as portrayed by anthropologists, masks a number of farming types, many sodalities, and gender differentiation. Ford's paper illustrates a departure from what he noted as a social and gender bias in ethnoscience literature based on the need of intellectuals well-versed in a language to provide encompassing answers about classifications of natural phenomena. It should be noted, further, that this significant departure was engendered by the need to put ethnoscientific research to the service of the community.

A Parallax Recognized: Refocusing Ethnoecology's Gaze

In an essay entitled "Cultural Parallax in Viewing North American Habitats," Gary Nabhan pointed out that:

> Despite such diversity within and between North American cultures, it is quite common to read statements implying a uniform "American Indian view of nature"—as if all the diverse cultural relations with particular habitats on the continent can be swept under an all-encompassing rug. Whether one is prejudiced toward the notion of Native Americans as extirpators of species or assumes that most have been negligible or respectful harvesters, there is a shared assumption that all Native Americans have used the flora and fauna in the same ways. This assumption is both erroneous and counterproductive in that it undermines any respect for cultural diversity. It does not grant cultures—indigenous or otherwise—the capacity to evolve, to diverge from one another, to learn about their local environments through time. (1995:91)

Nabhan used a visual metaphor in analyzing this problem, referring to the definition of parallax as it is employed in photography and astronomy as "the apparent displacement of an observed object due to the difference between two points of view."

In my paper, I investigate the various angles by which the reality of the Manupali watershed in southern Philippines may be refracted for people belonging to different ethnic, gender, and age groups. Adapting a method traditionally used in clinical psychology, the Thematic Apperception Test, to access the culturally relevant indicators of sustainability and quality of life, I discovered culturally shared dominant themes such as usefulness, diversity, and beauty of the environment and its various features that are more salient for the local population than economic value or commercialization potential in and of itself. On the other hand, certain trends of intergroup variation were demonstrated, indicating that lenses and latitudes are heavily influenced by the position of actors in the internal differentiation of society. Finally, based on the results, a displacement, or parallax, is apparent between quantitative, operational indicators of sustainability or end-states in development planning and the indicators that the local community, or sectors thereof, consider relevant and significant.

Devon Peña's paper focuses even more sharply on the dissonance in perception and decision making along the lines of minority/majority and Western-scientific/folk. Peña presents the case of Chicano settlers in the San Luis Valley of southern Colorado who have evolved a cultural landscape that links the watershed ecosystems and the high mountain peaks with farming systems in the commons *(ejidos)* of the Spanish-Mexican land grants. Particularly illustrative is the difference between state hydrologists and extension agents, on the one hand, and Chicano farmers, on the other, in viewing the local system of irrigation, the acequia. According to Peña, the former, along with some environmentalists, criticize the acequia system for being wasteful because the earthen ditches leak water. In contrast, the latter see water seepage as a boon for a variety of trees and shrubs that grow along the earthen ditches, providing a natural biological corridor that enhances biodiversity as well as creating a windbreak that prevents soil erosion.

The parallax between two "different visions" regarding plant genetic resources conservation is the subject of Daniela Soleri's and Steven Smith's paper. The authors argue that plant genetic resources conservation by the formal plant-breeding sector is really geared toward the preservation of folk varieties as exotic sources of raw materials for crop improvement and as weapons in the "relay race" of keeping modern agriculture slightly ahead of the rapid development of pest and pathogen resistance. By comparison, from the informal, farmer/breeder perspective, what counts is not so much the allelic diversity or the presence or absence of specific traits but rather "the combination of both local adaptation to the often difficult growing conditions and the presence of culturally valued traits such as seed or cob color that can make some folk varieties important cultural symbols that contribute to the maintenance of social relations within and between communities." Soleri and Smith, using a very different form of argument, nonetheless make a point similar to that made by Dove and, in a slightly different sense, Peña: that conservation cannot be abstracted from local use and from cultural values that inform choices made by the people whose lives and livelihood are inextricably linked to these resources.

Negotiating the Commons: Ethnoecology Makes a Difference

The commons is a recurring theme in this volume. Pastoral lands, watersheds, forests and forest margins, and traditional fishing grounds have been governed by common property regimes that have tended to emphasize group survival over self-interest and long-term sustainability over short-term gain. These values, however, are presently being strained by population pressures and the expanding global market economy. Timothy Johns suggests that for populations that have lived in close harmony with the natural resources of the commons, "the disruption of inherent homeostasis in their way of life can have negative consequences. In addition to undernutrition and increased infectious disease, indigenous peoples undergoing a process of change seem particularly vulnerable to diseases of affluence, perhaps a reflection of their finely tuned relationship with phytochemicals and other aspects of the environment."

Citing the pastoral Maasai of East Africa, who supplement their rich, dairy-based diet with bark and roots that have cholesterol-lowering and antioxidant properties and whose ethnobotany has only one word for "tree" and "medicine," Johns predicts that development coupled with diminishing access to commons resources for indigenous peoples will affect not only their nutrition and health but also their very sense of self. "Without cattle, a Maasai is not a Maasai," according to Johns, and for that reason alone there is need to look closely at substituting a private property framework for common property regimes.

A different take is explored by James Affolter and Marta Lagrotteria with reference to wildcrafting of medicinal and aromatic herbs. They discuss the case of the province of Córdoba in Argentina, where, because of the national demand for tea and other herbs, harvesting is done for most of the year, sometimes into the winter. This practice has led to a self-perpetuating cycle of economic hardship for the families of primary collectors and a degradation of the resource base caused by the lack of any real incentive to regulate harvest at sustainable rates. The issue, they point out, is "the risk of harvesting at nonsustainable rates by individuals motivated largely by self-interest and, even more basically, by self-preservation (which) raises the possibility of a 'tragedy' not dissimilar to that caused by overgrazed pastoral lands." The authors discuss a multisectoral initiative to domesticate the most commercially important and ecologically threatened herbs in order to spare the commons from overextraction. This raises several important questions that demand a rethinking of the common property concept vis-à-vis agricultural production, sustainable harvesting, and marketing of botanical genetic resources.

Scott Atran examines the problem of commons management in Mayaland from the perspective of cognitive models, emphasizing the centrality of the role played by information distribution and communication networks. According to Atran, a study of knowledge structures opens the possibility that there are cognitive means somewhat independent of institutional aspects for managing the commons. While conventional models of cooperation in commons management have assumed that actors uniformly share local knowledge, Atran argues for the need to factor in crucial asymmetries in local knowledge—emanating, in the case of the forests of lowland Mexico and Guatemala, from a convergence of local Maya, recent immigrants, and nongovernmental organizations (NGO) as well as intragroup distribution of expertise and gender-based specialization. The cultural consensus model should be used in this regard not only to identify foci of shared understandings but also—and perhaps more important—to anticipate systematic differences in the face of continual transmission and negotiation of meanings among various actors.

Ethnoecology's Relevance: Local Knowledge in Global Context

The native cultigens and cultural legacy of local peoples have been—and, in some cases, still are—considered a part of the common human heritage that should benefit everyone through unrestricted sharing. The problem, as we are slowly and sadly discovering, is that landraces and knowledge are "commodities" that are not exempt from the political economy of resource extraction in which resources are funneled to the developed First World without commensurate returns to the less developed Third World. In the context of increasingly strict plant protection laws and plant breeders' rights in developed countries, plant genetic resources and intellectual property rights activists—interestingly enough, from both worlds—have characterized the lopsided situation from the point of view of developing countries as one in which "what is mine is ours, but what is yours is yours."

According to Darrell Posey, intellectual property rights (IPR) is a Western concept that evolved out of a need to protect, through a system of legal sanctions and rewards, individual, technological, and industrial innovations. As such, this concept is not wholly compatible with indigenous or local systems of thought. He suggests that traditional resource rights, or TRR, which have less of a Western bias, may be a more appropriate mechanism of protection than the "limited and limiting" concept of IPR. Reflecting on one of the major themes of this volume, the need to expand our concerns from categories to strategies, Posey argues that "the critical factor here is to link folklore and plant genetic resources with intellectual property . . . (a) complicated legal linkage that allows for the expansion of the concept of IPR to include traditional knowledge not only about species use but also about species management." With this urgent task in mind, Posey poses a question: "If the 'new science of dialogue' cannot be developed out of ethnoecology, where can it come from?"

David Stephenson maintains that it is premature to abandon IPR tools for two fundamental reasons: First, Western legal tools have significantly evolved in the last two decades, giving rise to adaptations that may open more flexible IPR mechanisms that are more appropriate for indigenous peoples. Second, the difference between indigenous peoples and modern corporations may be overstated. According to Stephenson, "Indigenous peoples have knowledge, familiarity with local natural resources, and . . . creative expressions of their cultural experiences that can give an important competitive commercial advantage in the world marketplace." Unfortunately, like some modern companies, some indigenous groups do not realize the value of what they have until others have appropriated it for their own gain, as in the case of the registered trademark "Hopi blue," which currently poses a legal threat to any Hopi who may wish to market products under the name "Hopi." Stephenson argues that it would be a mistake to dismiss IPR considerations for indigenous peoples given the present power structure, inasmuch as "it is precisely because they often have inferior political status that a search for protecting their intellectual and cultural property is even more urgent." Anthropologists can play a vital role in negotiations to formulate culturally appropriate mechanisms of legal protection for intellectual property of indigenous groups.

Native healers, and the knowledge they possess regarding the pharmacological value of tropical species of plants, constitute an invaluable resource for all humanity. Yet, as Katy Moran, points out, virtually no compensation has been returned to indigenous communities for their intellectual contribution to the drug discovery process—a process that generates billions of dollars in revenues annually for developed countries. Moran's paper describes efforts to help formulate a strategy that promotes the conservation of tropical forests, particularly medicinal plants, while at the same time compensating indigenous peoples for their intellectual property, particularly their traditions of using medicinal plants. Moran stresses the need to strengthen the capacity of local cultures to monitor and manage the process. The "Medicine Women" project, for example, imparts to local women herbarium skills as well as the capability to assay plants on site. In the end, Moran argues, "equity means not only compensation but also and more important equal standing among participants in making decisions about what form compensation should take."

The intimate relationship and interdependence between local people and their environment developed a deep sense of the responsibility humans have to nature as well as to each other. However, as Christine Kabuye laments, population pressure and modern lifestyles have derailed many of these cultural priorities. Some experts and entrepreneurs have taken advantage of the free access regime to attach intellectual property protection to products emanating from traditional resources and knowledge, and to require payment for their use from their very source. Kabuye argues that "while there is a lot in bioprospecting to benefit humanity, there is no reason why the benefits should be one-sided." Big brother should not seek to profit from the trusting nature and relative inexperience of little brother in the global marketplace, but instead should demonstrate true concern for indigenous communities and humanity's pool of natural resources.

Of Acequias and Sweet Potato Vines: Postparadigmatic Meanderings

As acequias weave their way across ejidos, leaking life-sustaining water droplets that invite a multitude of plants to take root; as sweet potato vines climb and twine over hedges, quietly but efficiently covering everything in their path and creating miniature landscapes; as hogans continue to locate the Diné in space and time; and as humans defend their acequias, sweet potatoes, and hogans from obliteration, our metatheories of classification are rendered increasingly artificial and inadequate. Ethnoecology faces a challenge—to loosen up its rigid categories, to seek its meaning and validation not in models and artifacts characterized by perfect symmetry, correspondence, and consensus but in its ability to document, analyze, and articulate the variation, both inter- and intracultural, of the distribution of knowledge and interests that orient decision making and practice within any given environment. To paraphrase

Jean-Francoise Lyotard's injunction (1984), we need to calm our fear of differences, and develop our tolerance for the incommensurable.

Exciting frontiers open up as we explore traditional systems of water management among the Zuni and the effect of grand irrigation schemes; the conservation of archaic cultigens in contemporary agriculture; the differences between informal farmer/breeders and formal plant breeders in their strategies and their goals; the role of masticants and plant nutrients in traditional diets and how they are affected by changes in the management of the commons; and the clash as well as points of negotiation between Western systems of intellectual property rights and traditional resource rights. Many of the interesting questions, as we can see, are at the interface between the local and the global—in fragmented, contested, and situated ethnoecologies—and it is for this reason that ethnoecology needs to take heed of the highly, if at times subtly, politicized context of cognition and behavior.

George Marcus and Michael Fischer, in *Anthropology as Cultural Critique* (1986:8), observed that "social thought in the years since (the 1960s) has grown more suspicious of the ability of encompassing paradigms to ask the right questions, let alone provide the right answers, about the variety of local responses to the operation of global systems, which are not understood as certainly as they were once thought to be under the regime of 'grand theory' styles." Moreover, according to Worster (1993), disturbance, flux, and disorder have displaced stability, equilibrium, and order in holding sway over ecological thinking, a development not isolated from the chaos theory that has revolutionized physics and chemistry. How will ethnoecology, as an approach within anthropology that is (1) implicated in grand synthesis as part of "the new ethnography" and (2) in constant interaction with the natural and physical sciences, be affected by this radical reorientation? What does ethnoecology have to contribute to the epistemological and methodological refashioning that current self-critical reflection calls for? Finally, what role can ethnoecology play in interdisciplinary dialogue and action outside of anthropology? In short, what is its relevance in today's world?

I end this introduction to the volume with some preliminary suggestions that derive inspiration from tenacious acequias and sweet potato vines. One is that ethnoecology should maximize its comparative advantage, which, I would argue, lies more in its methodology than any once-and-for-all theory. I refer to this as a comparative advantage because ethnoecology, with its rich and varied repertoire of methods rooted in rigorous fieldwork, does not have to be confined to the analysis of narratives. It can infuse the whole issue of representation with a new perspective by means of skillful crafting of combinations of methods that allow our former "subjects" and informants to represent themselves or, at the very least, to help us countercheck our efforts to articulate their voices. It can draw on a long, undisputed tradition of putting local knowledge at center stage and emphasizing the significance of cognition to behavioral outcomes. Second, ethnoecology needs to come to terms with the situated nature of knowledge, the constraining as well as liberating effect of this locatedness, and the importance of history, power, and stake in shaping environmental perception, management, and negotiation. I believe that this is potentially where ethnoecology can make the greatest contribution to interdisciplinary research and even to advocacy in such areas as conservation, sustainability, and equity, because no other approach can draw on a jeweler's tool kit that is so promising for illuminating nuances and dimensions that more operational, quantitative, and macro approaches tend to neglect or gloss over.

The meanderings of the acequias and sweet potato vines belie an obstinate rootedness, poke fun at rigid linearity, and symbolize an almost compulsive drive to link up and connect. I would wager that implicit in these properties is an attraction for what lies below the surface, an impatience for narrow confines, and a tolerance for surprises. As I think the papers in this volume amply illustrate, ethnoecology cannot do too badly by going in this direction, for knowledge's—and knowledge bearers'—sake.

References

Banks, William P., and David Krajicek. 1991. "Perception." *Annual Review of Psychology* 42:305–31.

Berlin, B. 1992. *Ethnobiological Classification: Principles of Categorization of Plants and Animals in Traditional Societies.* Princeton: Princeton University Press.

Berlin, B., D. E. Breedlove, and P. H. Raven. 1974. *Principles of Tzeltal Plant Classification.* New York and London: Academic Press.

Bourdieu, P. 1987. "What Makes a Social Class?: On the Theoretical and Practical Existence of Groups." *Berkeley Journal of Sociology: A Critical Review* 31:1–18.

Conklin, H. 1954. "The Relation of Hanunuo Culture to the Plant World." Ph.D. diss. Yale University.

———. 1961. "The Study of Shifting Cultivation." *Current Anthropology* 2:27–61.

Ellen, R. F. 1993. *The Cultural Relations of Classification: An Analysis of Nuaulu Animal Categories from Central Seram.* Cambridge: Cambridge University Press.

Ford, R. I. 1978. "Ethnobotany: Historical Diversity and Synthesis." In *The Nature and Status of Ethnobotany.* R. I. Ford, ed. Anthropological Papers, Museum of

Anthropology, University of Michigan, no. 67, pp. 33–49. Ann Arbor: University of Michigan Press.

Fowler, C. 1977. "Ethnoecology." In *Ecological Anthropology.* D. Hardesty, ed. New York: Wiley.

Frake, C. O. 1962. "Cultural Ecology and Ethnography." *American Anthropologist* 64: 53–59.

Goodenough, W. H. 1957. "Cultural Anthropology and Linguistics." In *Report of the Seventh Annual Table Meeting on Linguistics and Language Study.* P. L. Garvin, ed. Washington, D.C.: Georgetown University Monograph Series on Languages and Linguistics.

Hunn, E. 1977. *Tzeltal Folk Zoology: The Classification of Discontinuities in Nature.* New York: Academic Press.

———. 1982. "The Utilitarian in Folk Biological Classification." *American Anthropologist* 84:830–47.

———. 1989. "Ethnoecology: The Relevance of Cognitive Anthropology for Human Ecology." In *The Relevance of Culture.* N. Freilich, ed. New York: Bergin and Garvey.

Lease, G. 1995. "Nature Under Fire." In *Reinventing Nature? Responses to Postmodern Reconstruction.* M. Soule and G. Lease, eds. Washington, D.C.: Island Press.

Lévi-Strauss, C. 1966. *The Savage Mind.* London: Weindenfeld and Nicolson.

Lyotard, J. F. 1984. *The Postmodern Condition: A Report on Knowledge.* Minneapolis: University of Minnesota Press.

Marcus, G., and M. J. Fischer. 1986. *Anthropology as Cultural Critique: An Experimental Moment in the Human Sciences.* Chicago and London: University of Chicago Press.

Meinig, D. W. 1979. "The Beholding Eye: Ten Versions of the Same Scene." In *The Interpretation of Ordinary Landscape.* Oxford: Oxford University Press.

Nabhan, G. 1995. "Cultural Parallax in Viewing North American Habitats." In *Reinventing Nature? Responses to Postmodern Deconstruction.* M. Soule and G. Lease, eds. Washington, D.C.: Island Press.

Randall, R. A., and E. Hunn. 1984. "Do Life Forms Evolve or Do Uses for Life? Some Doubts about Brown's Universals Hypotheses." *American Ethnologist* 4: 1–26.

Romney, A. K., and R. G. D'Anrade. 1964. "Cognitive Aspects of English Kin Terms." *American Anthropologist* 66:146–70.

Sturtevant, W. C. 1964. "Studies in Ethnoscience. In Transcultural Studies. In Cognition." *American Anthropologist* 66:99–113.

Toledo, V. 1992. "What is Ethnoecology? Origins, Scope, and Implications of a Rising Discipline." *Etnologica* 1(1): 5–21.

Worster, Donald. 1993. "Organic, Economic, and Chaotic Ecology." In *Major Problems in American Environmental History.* Carolyn Merchant, ed. Lexington, Mass.: D.C. Heath and Company.

PART I

Ethnoecology

Cultural Memory and Sense of Place

2

The Value of Subsistence for the Future of the World

EUGENE S. HUNN

Documenting traditional ecological knowledge has become a growing concern in recent years. The acronym TEK, which stands for "traditional environmental knowledge" (Williams and Baines 1993), is popular, as is IK, "indigenous knowledge" (Brokensha et al. 1980; Cunningham 1991; Warren et al. 1991); others prefer the label "local environmental knowledge." More recently "ethnoecology" has been emphasized (Toledo 1992). The profusion of acronyms and competing headings suggests the rapid emergence of this perspective from within several independent academic networks.

Such bodies of traditional knowledge are gravely threatened, in imminent danger of going to the grave with the present generation of elders. One could play devil's advocate and argue that today's world is a very different world than that with which these elders learned to deal; that their traditional knowledge is thus obsolete; and that new sorts of knowledge will better serve present and future generations. In particular, it might be asserted that Western scientific knowledge in such fields as biochemistry, global biogeography, and evolutionary ecology will suffice to guarantee our future. Given that cultural knowledge is dynamic and must change in response to changing requirements for survival and success, how can we justify devoting substantial time, effort, and resources to the task of preserving traditional environmental knowledge?

Three Reasons to Preserve TEK

TEK is both local and fragile. That TEK is local rather than global in scope is a consequence of the context of its acquisition, transmission, and use. It is acquired via direct personal experience, is transmitted orally within a community, and is validated by its relevance to the daily struggle to wrest a livelihood from one's land.

It is fragile because it is local. Knowledge common to one community is specific to its immediate environment and will not be shared widely in other communities. Thus that particular body of knowledge lives and dies with the community that sustains it, and that it in turn sustains. A corollary is that the value of TEK is additive across the world's cultures. Nevertheless, many formal characteristics of such cultural knowledge systems may be widespread or universal, reflecting the psychic and experiential unities of humankind (Berlin 1992). Thus, preserving even the essential features of a few such systems may inform us deeply about our common humanity.

By contrast, Western science strives for universal relevance and global scope. Western scientific research findings are published, recorded in a form that is both permanent and accessible to the scrutiny of any person in the world with the means and motive to consult that record. The local nature of TEK is both a weakness and a strength. Local knowledge systems are less likely than global systems to support powerfully general theories. However, local environmental knowledge systems have proved in many cases to provide a description of local environments superior in detail and coherence to that of Western biological science (Diamond 1966; Hunn and French 1981; Johannes 1981; Jones and Konner 1976; Nations and Nigh 1980). Such systems are grounded in lifetimes of intimate daily observation, a luxury not available to the vast majority of professional Western biologists.

If it be granted that TEK is fundamentally sound as science (Hunn 1993), then TEK complements the findings of Western science rather than being superseded by them. In that case, it is well worth the effort to preserve such systems as far as is now possible as part of the published scientific record.

So far, our justification for preserving TEK has focused on the value of the information such systems of knowledge may contain; how they may augment the corpus of available scientific data. That certainly is the thrust of the popular promotion of ethnobotany by Mark Plotkin (1993) and others (such as Arvigo and Balick 1993; Cox and Balick 1994; and Schultes 1990). From this perspective, TEK should be preserved as a potential source of information that may lead to the development of new cancer treatments, new disease-resistant crop varieties, or renewable and biodegradable substitutes for the materials currently required to drive our industrial technologies (Head and Heinzman 1990; Nabhan 1985, 1989). In such cases TEK is but one input into a modern scientific process intended to prop up the contemporary global status quo. Ethical concerns have with good reason been raised about such "technology transfers" in reverse (Martin 1995:239–46; Posey 1983) as exploiting TEK to support a global system that is deeply implicated in its destruction. It can be argued that we are destroying the very communities that created the traditional environmental knowledge we now seek to preserve in our libraries and archives.

I believe that there are more compelling reasons to preserve TEK than the contribution of the knowledge gained thereby to the advance of Western cultural enterprises. First, TEK is a monument to our common humanity. Meticulous descriptions and comparative analyses have shown that all cultures produce scientific knowledge (Berlin 1992). In short, we are all scientists, at least part of the time. The evidence of TEK forces us to see ourselves and our science in a different light. No longer can we take refuge behind the myth of the superiority of Western civilization as the source of all science. (Nor must we take all the blame for it, either). Furthermore, the evidence of TEK exposes the flawed logic of those who argue that science can or should be value free (Feyerabend 1987). For TEK is inextricably embedded in systems of moral value and integrated with the global meanings we call "religion."

We may profitably consider "animism" in this regard. Animism is a religious principle upheld by many traditional subsistence-oriented communities of hunter-gatherers, fisher folk, and horticulturists (Brightman 1993; Feit 1973; Hunn 1990; Nelson 1982). As an explanatory theory, animism postulates that all living things (often including as well "nonanimate" natural elements such as wind, water, and stone) are animated by spirit, which entails a mode of consciousness, intelligence, will, and memory comparable to that attributed to human beings. As committed Darwinists, we consider such an explanatory principle to be quite simply false because it is anthropomorphic. On the contrary, we believe ["know"] that the natural world (perhaps excepting human action) is governed by impersonal, mechanistic forces and reflects an accidental design. However, if we are honest scientists we should be willing to admit that our belief in those impersonal, mechanistic forces is based on much faith combined with a lack of convincing contradictory evidence in our experience. The same test sustains animists' faith that the elements of their universe are conscious, moral entities. Such a belief has effectively guided them in their interactions with nature "since time immemorial." Why would they doubt it? In short, we should study TEK carefully in its social and cultural context to escape a perhaps fatal blind faith in our own particular brand of scientific truth (Feyerabend 1987:20).

There is yet another reason to preserve TEK: as designs for independent alternatives to the globalization of a market mentality that at present comes close to overwhelming all competition for the hearts and minds of humanity. The socialist alternative—the so-called Second World, which complemented the First and Third Worlds of world systems theory—has all but collapsed. I suspect that the fate of the socialist alternative demonstrates that it was no real alternative to the industrial mode of production and the global market system that has forced the adoption of so-called economies of scale in our every productive endeavor. Communities that produce the TEK we seek to preserve are Fourth World communities, encapsulated within nation states, invariably small in population and poor by our standards of surplus wealth produced for market exchange. These communities are tied to very specific places that constitute their habitat and that are the target of their environmental knowledge. These "primitive" communities at the margins were abandoned by human history with the rise of states and of markets for labor and land (Marx 1964).

Yet some have persisted; some have resisted this historical trend. But their continued existence as partially autonomous and relatively self-sufficient communities within modern nations has not been tolerated graciously by proponents of progress. Rather, from the dominant development paradigm, such "backward" communities occupy valuable lands and constitute a pool of labor that could be integrated and employed "more productively" in furthering the reach of the world market.

Perhaps these ways of life, their religious visions, and their fragile, local systems of environmental knowledge are doomed. We may preserve a record of their ways of life, much as we might preserve some genetic traces of an obscure landrace of tepary bean, or corn, or potato in a vial of liquid nitrogen in a genetic resource bank for analysis by a future generation of scientists (Nabhan 1989), perhaps wiser and less rushed than our own.

I would like to argue that there is more at stake here than a historical record of past human accomplishments. I believe that our future as a species may hinge on preserving in print not just traces of traditional ecological adaptations, expressed as TEK, for the contemplation of future scholars. Rather, TEK systems embody the cultural diversity of the human species. As such their role in the evolutionary future of our species may be compared to the role of biodiversity in the future of life on earth. TEK bears more than a fanciful resemblance to the genome of a species (Boyd and Richerson 1985; Dawkins 1976; Pulliam and Dunford 1980; Ruyle 1973). That genome is a blueprint for a way of life that has survived. Each gene of the genome is a bit of information essential to the manifestation of the species in the life of each individual.

Likewise, TEK is a stock of ideas essential to the expression of a culture in the lives of its constituent individuals. The extinction of a species has been likened to burning a library. Might we make the same claim with regard to the extinction of a way of life and the loss of TEK that ensues? In each case extinction limits future evolutionary options. In the cultural case, we may be left with just one—the global capitalist consumer society.

I believe we should not only preserve a record of lost TEK in libraries and archives but also strive to preserve systems of TEK in vivo, in situ as radical alternatives to the present world system. To return to an earlier point: TEK is grounded in daily life, is sustained by its relevance to the exigencies of making a living off the land (Hunn 1982). That is why TEK is everywhere endangered today: Traditional communities rooted in ancestral lands have everywhere been dispossessed of those lands and in the process have been alienated from the work of harvesting the resources of their lands. Traditional knowledge of the land and of its resources is no longer relevant to the survival of the present generation as they pursue new livelihoods as migrant workers, factory employees, or panhandlers. The phrase "use it or lose it" well describes this situation. To preserve the full value of TEK, we must allow the members of traditional communities the opportunity to apply it in their daily lives, to maintain it, modify it, and pass it on to their descendants as still useful knowledge.

I am convinced that there is some measure of hope that the communities that gave us TEK can find space in which they may continue to exist, practicing their own ways of life as a distinct alternative to our own. The continued existence of such communities represents choices for our future. That some people, if given the chance, will choose such alternatives over participation in our vaunted modern world is proved by recent developments such as the "outback" movement among Australian aborigines (Coombs et al. 1989), the persistence of reservation-based tribal communities in the United States in the face of confident predictions of their speedy demise, the demand by Native Alaskans and more recent immigrants to that vast state of a right to practice a subsistence way of life (Berger 1985), and a resurgence of interest in traditional farming and fishing practices in Third World nations such as Mexico (Toledo et al. 1985) and Palau (Johannes 1981:74–75).

Lest I be dismissed as a romantic, let me examine in more detail two cases in which "customary and traditional" subsistence rights have been vigorously and, by and large, successfully defended in the face of powerful political and economic opposition. Perhaps best known is the case of treaty guarantees of rights to fish, hunt, and gather shellfish, roots, and berries throughout traditional territories ceded by treaty in the Pacific Northwest (Cohen 1986). An exemplary statement of such guarantees on which contemporary treaty rights are based is the following provision of the "Treaty between the United States and the Yakama Nation of Indians" negotiated in 1855 by Isaac Ingalls Stevens, the first territorial governor and Indian agent for Washington Territory: "The exclusive right of taking fish in all the streams, where running through or bordering said reservation, is further secured to said confederated tribes and bands of Indians, as also the right of taking fish at all usual and accustomed places, in common with citizens of the Territory . . . ; together with the privilege of hunting, gathering roots and berries, and pasturing their horses and cattle upon open and unclaimed land" (Treaty of Medicine Creek, 10 Stat., 1132, Article III, paragraph 2).

This treaty language and its underlying intent have been interpreted many times by federal courts up to and including the U.S. Supreme Court (Cohen 1986:107–17). There is a broad legal consensus that the descendants of the indigenous "tribes," though they must accommodate their ways of life to the presence of the Euroamerican colonists (retaining only the right of harvest in common with [other] citizens), nevertheless are not required to abandon those traditions entirely. In fact, they are guaranteed the means, both legal and ecological, of continuing to support themselves by harvesting traditional resources at "all usual and accustomed places."

The famed Boldt decision (*United States v. Washington*, 384 F. Supp. 312 [1974]) of Justice George Boldt defined "in common" as implying a 50-50 division of available resources between tribal harvests and those of non- Indian sport and commercial interests. This decision further directed that the tribal right should not be rendered meaningless by virtue of destruction of the habitat necessary to sustain the resources in question (Cohen 1986:137–53). In what is known as Phase II of the Boldt decision, the question of how to protect the habitats essential to preserve the tribes' subsistence rights is addressed. In sum, our legal system opens a space for an alternative relationship to exist between a community of people and their local environment.

Cynics see the tribes as just another user group demanding their market share of a fast-dwindling resource. In my experience, there is reason to believe otherwise. Many tribal members oppose developing reservation resources for maximum profit, arguing that they should be preserved as a sacred trust or to preserve the solitude of the place or the purity of its water or the abundance of roots, berries, and game. Such "traditional" values may not be incorruptible, but they have withstood great pressure and survive to the present day.

My second example is the subsistence provisions of the Alaska National Interest Lands Conservation Act (ANILCA, Public Law 96-487; 1980). ANILCA added approximately forty-five million acres to the National Park System. In sharp contrast to the legal mandate for national parks in the lower forty-eight states, ANILCA provided for the continuation of subsistence uses. These were defined as: "the customary and traditional uses by rural Alaska residents of wild, renewable resources for direct personal or family consumption as food, clothing, tools or transportation; for the making and selling of handicrafts out of nonedible by-products of fish and wildlife resources taken for personal or family consumption, and for customary trade." Specifically, the law directed the governmental agencies responsible for managing federal lands in Alaska to give priority over other consumptive uses to subsistence uses of natural resources, stating that "the continuation of the opportunity for subsistence uses by rural residents of Alaska, including both Natives and non- Natives . . . is essential to Native physical, economic, traditional, and cultural existence and to non-Native physical, economic, traditional, and social existence" (section 801).[1] (I'm unclear why it is essential for Native "cultural" existence but for non-Native "social" existence.)

National Parks in the lower forty-eight were established in many cases to preserve "wilderness," which effectively disallowed all "consumptive uses" of park resources, whether for commercial, sport, or subsistence purposes. Such park lands, instead, were "consumed" by hordes of tourists during brief vacations from the pressures of urban life. The notion that National Parks could accommodate ongoing subsistence harvesting—such as hunting, fishing, trapping, and wood cutting—by local "rural residents" was at first difficult for Alaska Park Service personnel to grasp. However, a "subsistence life style" is what attracted many early settlers to Alaska in the first place and is a political force to be reckoned with in Alaska today.

Subsistence is yet more sacrosanct in the dozens of Alaskan Native Indian, Aleut, and Eskimo communities that resist assimilation into the American mainstream. For many rural residents, their annual harvest of moose, caribou, or salmon and access to wood for house logs and fuel are an economic necessity. For others, the cost of purchasing, maintaining, and operating "subsistence tools" such as snow machines, outboard motors, and chainsaws may nearly balance the dollar value of their subsistence harvests, requiring that they work for wages to support their "subsistence habit." In either case, the opportunity to engage nature in this direct way gives meaning to their lives, strengthens their family and community ties, and stands ever in the way of more capital-intensive land development schemes that might deprive them of their subsistence privileges.

The National Park Service has embarked upon a substantial research program since 1989 to document the realities of subsistence for communities within their jurisdiction. I was involved in one such study, which investigated the significance of the subsistence harvesting of plants for six rural communities located near Lake Clark National Park and Preserve, southwest of Anchorage (Johnson et al. 1997). These six communities differed in residential history, ethnic composition (Dena'ina Athabaskan Indian, Yup'ik Eskimo, and Euroamerican), religion (for example, Russian Orthodox, Evangelical Protestant), and degree of involvement in the cash economy (although extensive in all cases). Of particular significance to the degree of commitment to a subsistence way of life was the "rootedness" of the community in its local environment—which we measured in terms of the likelihood that household heads and their parents were born in the same community or region—and the complexity of the linkages between the subsistence practices of individual households and the social life of the community.

Several conclusions seem justified on the basis of this study: (1) subsistence should be understood as a long-term relationship between a community and its land and resource base, rather than as a strictly economic activity; (2) subsistence is dynamic, rooted in past practices but of necessity adapting to technological, demographic, economic, social, and political changes; (3) subsistence activities are integral to the life of families and communities, an aspect of their identity and continuity expressed in subsistence work; and (4) the meaning of subsistence is different for each community, varying with the ethnic, religious, and economic histories of communities and their component families. Finally, effective management of subsistence activities on federal lands demands that a truly cooperative spirit pervade all aspects of management, including monitoring, policy formation, and enforcement.

As in Alaska, Puget Sound Indian fishermen exercise their treaty right to fish at "all usual and accustomed grounds and stations" by means of rather untraditional technology. They no longer use hand-hewn cedar dugouts, fishing lines and nets of twined nettle bark bast, or harpoon blades of mussel shell fused to fir spear shafts with spruce pitch. Instead, they set nylon nets from fiberglass skiffs powered by eighty-horse Merc outboards. Some observers of this scene find it hypocritical of the Indians to assert a customary right to harvest traditional resources using such modern, "white man's" methods. To my way of thinking, there is no necessary contradiction. Technologies have always evolved to meet new demands and to take advantage of new means. What is critical is not the means but the motives.

Are these treaty fishermen engaged in an industrial and commercial enterprise the goal of which is to maximize profits and to compete for an ever-greater market share? Or are they engaged in making a living by harvesting the resources of their ancestral homeland in hopes that their children and grandchildren may do the same? The issue of production for direct consumption—that is, do they eat all and only the food they harvest?—versus production for exchange—do they sell some of what they harvest for money to buy necessities that they do not or cannot produce locally?—clouds the issue somewhat. However, the notion of an entirely self-sufficient "primitive" community does not match ethnographic or historical reality. Some production for exchange is reported for most if not all subsistence-based communities. At what point production passes that invisible line differentiating subsistence from commercial production is an issue I leave for future argument. Still, the analytical distinction between subsistence and market orientations is critical to my argument here. TEK is a consequence of subsistence-based production. We cannot preserve the one without preserving the other.

Karl Marx, by Way of Conclusion

I have argued that to preserve TEK we need to encourage the continuity of subsistence-based communities where such knowledge is produced. This proposal may seem to be against history, which has conspired everywhere to destroy such communities. Such is Marx's vision of human history: Subsistence-based communities are survivals of Marx's "Archaic Formation," which everywhere preceded the development of societies structured by class divisions. Marx concluded, "The precondition for the continued existence of the [primitive] community is the maintenance of equality among its free, self-sustaining peasants, and their individual labour as the condition of the continued existence of their property" (1964:73). The growth of population and the reliance upon war by one community against another to gain land to support that population led to the emergence of states (Marx 1964:71; Leacock 1972:46–57). The emergence of a trader class in the "interstices" of these early states sowed the seeds of capitalism, which like a great weed overwhelmed the feudal garden.

Marx's utopian vision of a communist end point to history was predicated on his expectation "that communism would be a re-creation, on a higher level [of productivity], of the social virtues of primitive communalism . . ."

(Hobsbawm 1964:51). Marx at first welcomed "capitalism as an inhuman but progressive force," but in his maturity "found himself increasingly appalled by this inhumanity" (Hobsbawm 1964:50). He thus came to stress "the viability of the primitive commune, its powers of resistance to historical disintegration . . ." (p. 50).

The inhumanity Marx despised in capitalism was not, however, informed by a clear sense of capitalism's ecological impact, for no science of ecology existed in his day. He recognized that human production was ultimately dependent on nature: "The earth is the great laboratory, the arsenal which provides both the means and the material of labour, and also the location, the *basis* of the community . . . which produces and reproduces itself by living labour" (Marx 1964:69). However, his hope for a communist utopia presumed an infinite productive capacity. I share Marx's horror at the destructive power of global capitalism, its insatiable hunger for resources with which to turn a profit, "its incredible potential for both enormous creation and for insane—perhaps ultimate—destruction: the heritage of the 20th century" (Leacock 1972:57). But I see it as a threat not only to human values but also to the earth itself.

What were the moral strengths of the "primitive commune" that Marx saw being subverted by history? Essentially, he valued the organic unity of a community of human beings tied to their land by their own labor with which they produced their livelihood and in so doing reproduced their community. TEK constituted the intellectual capital for such communities and was in turn a product of the unity of land and labor that sustained it.

What has led to the disintegration of these primitive communities across the globe during the past five thousand years? Marx saw this as a complex question, for the archaic base assumed various forms, which he labeled the "Asiatic," "Slavonian," Germanic," and "ancient [Greek]" (Hobsbawm 1964:35), each with its peculiar evolutionary potential.

In keeping with his dialectical method, Marx sought the roots of this disintegration in the internal contradictions of the primitive communal social formation. What were the contradictions that contained the seeds of its destruction? Marx answered, "War . . . is the great all-embracing task, the great communal labor [of the primitive community], and it is required either for the occupation of the objective conditions for living existence [i.e., land] or for the protection and perpetuation of such occupation" (1964:71). Thus, each community must be prepared to wage war to protect that share of the earth—"the arsenal which provides both the means and the material of labour"—that sustains it. The consequence of war, however, is enslavement or enserfment of the conquered by the conquerors. Thus class divisions arise that split apart the community, the dominant class expropriating the labor of the subordinate class for its own purposes. War in turn was seen as a consequence of "the advance of population" (p. 83); "the mere increase in population constitutes an obstacle. If this is to be overcome, colonisation will develop and this necessitates wars of conquest. This leads to slavery. . . . Thus the preservation of the ancient community implies the destruction of the conditions upon which it rests" (pp. 92–93).

Yet the states and empires of antiquity remained, according to Marx, "system[s] of production for use," and thus, in such systems, "no boundless search for surplus labour arises from the nature of production itself" (quoted in Hobsbawm 1964:30). But the "ancient conception, in which man always appears . . . as the aim of production [is inverted by] the modern world, in which production is the aim of man and wealth is the aim of production" (Marx 1964: 84). Capitalism represents a radical departure from this "ancient conception." It emerges almost imperceptibly at first, taking root in the "interstices" of the ancient and feudal states: "Wealth as an end in itself appears only among a few trading peoples . . . who live in the *pores* [emphasis added] of the ancient world"; "The main agent of disintegration [of feudalism] was the growth of trade" (p. 64); "Crucial to the development of capitalism is therefore that of the world market" (Hobsbawm 1964:30).

Marx's analysis of human history is wrong on many details, and Marx is dead wrong, in my view, in his belief that capitalism is the penultimate stage in that history, to be followed by a utopian recreation of a global village sustained by the presumedly infinite productive capacity of humanity once liberated from class exploitation. Nevertheless, I am convinced by his argument that the root of our present predicament (as well as the source of our present precarious prosperity) lies in the mass alienation of human beings from the work of subsistence on their own land. My vision of the future is far more modest than Marx's. I believe that the present global market system, like that of the monolithic edifice of feudalism before it, is cracked and creviced. The "interstices" of the system may shelter alternative modes of production. The stubborn persistence of contemporary subsistence-based communities—like weeds that push up through cracks in the pavement—sustains that belief. Such communities survive by taking advantage of residual legal claims to control land that they then work with their own labor and resources, nurturing and protecting it for their children and grandchildren. They hold the world market and the national governments that serve it at bay. We should lend them our support, as their survival serves us all by preserving in the practice of their daily lives living examples that alternatives exist to the present world order.

Note

1. This section draws from a work plan entitled "A Critical Review of the Literature Associated with the Cooperative Management of Parks and Equivalent Preserves and Selected Case Studies Pertinent to the Management of Alaska Units of the National Park System," submitted to the U.S. National Park Service, Alaska Regional Office, by E. Hunn, D. Johnson, C. Sander, and C. Sawin-Wilson on behalf of the Cooperative Park Studies Unit, College of Forest Resources, University of Washington, Seattle, 1993.

References

Arvigo, R., and M. Balick. 1993. *Rainforest Remedies: 100 Healing Herbs of Belize*. Twin Lakes: Lotus Press.

Berger, T. R. 1985. *Village Journey: the Report of the Alaska Native Review Commission*. New York: Hill and Wang.

Berlin, B. 1992. *Ethnobiological Classification: Principles of Categorization of Plants and Animals in Traditional Societies*. Princeton: Princeton University Press.

Boyd, R., and P. J. Richerson. 1985. *Culture and the Evolutionary Process*. Chicago: University of Chicago Press.

Brightman, R. A. 1993. *Grateful Prey: Rock Cree Human-Animal Relationships*. Berkeley: University of California Press.

Brokensha, D. W., D. M. Warren, and O. Werner, eds. 1980. *Indigenous Knowledge Systems and Development*. Lanham, Md.: University Press of America.

Cohen, F. G. 1986. *Treaties on Trial: The Continuing Controversy over Northwest Indian Fishing Rights*. Seattle: University of Washington Press.

Coombs, H. C., H. McCann, H. Ross, and N. M. Williams, eds. 1989. *Land of Promises: Aborigines and Development in the East Kimberley*. Canberra: Centre for Resources and Environmental Studies, Australian National University.

Cox, P. A., and M. J. Balick. 1994. "The Ethnobotanical Approach to Drug Discovery." *Scientific American* 270(6):82–87.

Cunningham, A. B. 1991. "Indigenous Knowledge and Biodiversity: Global Commons or Regional Heritage?" *Cultural Survival Quarterly* 15(3):4–8.

Dawkins, R. 1976. *The Selfish Gene*. Oxford: Oxford University Press.

Diamond, J. 1966. "Zoological Classification System of a Primitive People." *Science* 151:1102–4.

Feit, H. A. 1973. "Ethno-ecology of the Waswanipi Cree; Or, How Hunters Can

Manage Their Resources." In *Cultural Ecology*. B. Cox, ed. Toronto: McClelland and Stewart.

Feyerabend, P. 1987. *Farewell to Reason*. London: Verso.

Head, S., and R. Heinzman, eds. 1990. *Lessons of the Rainforest*. San Francisco: Sierra Club Books.

Hobsbawm, E. J. 1964. "Introduction." In *Pre-capitalist Economic Formations: Karl Marx*. New York: International Publishers.

Hunn, Eugene S. 1982. "The Utilitarian Factor in Folk Biological Classification." *American Anthropologist* 84:830–47.

———. 1990. *Nch'i-Wána "The Big River": Columbia River Indians and Their Land*. Seattle: University of Washington Press.

———. 1993. "What is Traditional Ecological Knowledge?" In *Traditional Ecological Knowledge: Wisdom for Sustainable Development*. N. M. Williams and G. Baines, eds. Canberra: Centre for Resources and Environmental Studies, Australian National University.

Hunn, Eugene S., and D. H. French. 1981. "*Lomatium*, a Key Resource for Columbia Plateau Subsistence." *Northwest Science* 55:87–94.

Johannes, R. E. 1981. *Words of the Lagoon: Fishing and Marine Lore in the Palau District of Micronesia*. Berkeley: University of California Press.

Johnson, D. R., E. Hunn, P. Russell, M. Vande Kamp, and E. Searles. 1997. "Subsistence Uses of Vegetal Resources in and around Lake Clark National Park and Preserve." Final review draft, Field Station for Protected Area Research, USGS/BRD/FRESC, College of Forest Resources, University of Washington, Seattle.

Jones, N. B., and M. J. Konner. 1976. "!Kung Knowledge of Animal Behavior (Or: The Proper Study of Mankind is Animals)." In *Kalahari Hunter-Gatherers: Studies of the !Kung San and Their Neighbors*. R. B. Lee and I. DeVore, eds. Cambridge: Harvard University Press.

Leacock, E. B. 1972. "Introduction." In *Bound with the Origin of the Family, Private Property and the State*. F. Engels, ed. New York: International Publishers.

Martin, G. J. 1995. *Ethnobotany: A "People and Plants" Conservation Manual*. London: Chapman & Hall.

Marx, K. 1964. *Pre-capitalist Economic Formations*. New York: International Publishers.

Nabhan, G. P. 1985. *Gathering the Desert*. Tucson: University of Arizona Press.

———. 1989. *Enduring Seeds: Native American Agriculture and Wild Plant Conservation*. San Francisco: North Point Press.

Nations, J. D., and R. B. Nigh. 1980. "The Evolutionary Potential of Lacandon Maya Sustained-Yield Tropical Forest Agriculture." *Journal of Anthropological Research* 36:1–30.

Nelson, R. K. 1982. "A Conservation Ethic and Environment: The Koyukon of Alaska." In *Resource Managers: North American and Australian Hunter-Gatherers*. N. M. Williams and E. S. Hunn, eds. Boulder: Westview Press.

Plotkin, M. J. 1993. *Tales of a Shaman's Apprentice: An Ethnobotanist Searches for New Medicines in the Amazon Rain Forest*. New York: Viking.

Posey, D. A. 1983. "Indigenous Ecological Knowledge and Development of the Amazon." In *The Dilemma of Amazonian Development*. E. Moran, ed. Boulder: Westview Press.
Pulliam, H. R. and C. Dunford. 1980. *Programmed to Learn: An Essay on the Evolution of Culture*. New York: Columbia University Press.
Ruyle, E. E. 1973. "Genetic and Cultural Pools: Some Suggestions for a Unified Theory of Biocultural Evolution." *Human Ecology* 1:201–15.
Schultes, R. E. 1990. *The Healing Forest*. Portland, Oreg.: Dioscorides Press.
Toledo, V. M. 1992. "What is Ethnoecology? Origins, Scope and Implications of a Rising Discipline." *Ethnoecológica* 1(1):5–21.
Toledo, V. M., J. Carabias, C. Mapes, and C. Toledo. 1985. *Ecología y autosuficiencia alimentaria*. México, D.F.: Siglo XXI.
Warren, D. M., L. J. Slikkerveer, and S. O. Titolola, eds. 1989. "Indigenous Knowledge Systems: Implications for Agriculture and International Development." *Studies in Technology and Social Change no. 11*. Ames: Iowa State University.
Williams, N. M., and G. Baines, eds. 1993. *Traditional Ecological Knowledge: Wisdom for Sustainable Development*. Canberra: Centre for Resources and Environmental Studies, Australian National University.
Wilson, E. O. 1992. *The Diversity of Life*. Cambridge: Belknap Press.

3

Practical and Religious Meanings of the Navajo Hogan

LILLIE LANE

My people—the Navajos—prefer to call ourselves "Diné" (people, or human beings), but we also accept the term "Navajo," which was a name that other neighboring tribes and early Spanish explorers called us. According to the most recent official tally, Navajos number about 300,000, living in the Four Corners region of Arizona, Utah, Colorado, and New Mexico. The Diné live in round structures called in English "hogans" but pronounced by my people as "hooghan." When I was young, my grandparents and parents lived intermittently in hogans, so I grew up in several of them. I left home for an education and have now not resided in a hogan for some thirty years. Upon my return to the reservation, however, my fondness for the traditional hogan was rekindled. I found that these structures are no longer as abundant as before because they are being replaced with modern homes. For me, hogans conjure the carefree times of my childhood, my grandparents, and our traditional lifestyle.

Throughout my travels on Diné bike'yah—Navajolands—I keep my eye out for hogans. When I do detect them, some that I see are lived in, others are abandoned, and still others sit out on the endless landscape or in clustered neighborhoods waiting to host a family or the next healing ceremony. To the outsider, hogans may seem crude structures because they lack modern conveniences such as running water, electricity, and plumbing. Nevertheless, the Navajo hogan is a structure that embodies the social, religious, and philosophical teachings of the Diné.

There are two kinds of Navajo hogan. The first is the fork-stick style of hogan, the *alchi adeeza*. The fork-stick style is made with cedar logs all standing upright. The logs are covered with bark, and mud is plastered on top.

These hogans are cone shaped, with an entranceway. According to Navajo oral traditions, when the Diné came into the present world (which is considered to be the fifth world), they moved into a *t'acheeh*, which is a small, fork-stick structure traditionally used for spiritual and physical cleansings. The alchi adeeza is a somewhat later version of the t'acheeh, but it is also considered an old, traditional style. It, too, is considered male in nature.

The other hogan is the *hooghan nimazi*—a round structure made of cedar logs, bark, and mud. The round hooghan nimazi is designated female. It is a dome-shaped mud structure, and there are many styles in which it can be built. According to Navajo oral traditions, the male and the female hogans were ascribed specifically to the Diné by the holy people who lived prior to the earth-surface-people.

There is a Navajo story that tells about the first fire, which would not kindle when the Diné first emerged into the fifth world. At the time they were using the t'acheeh as shelters, and they had ventured on building the alchi adeeza—the fork-stick style. There was at first much bewilderment and inquiry as to why the fire would not light. After much deliberation, the elder men, who may have been medicine men, realized that the fire in the new alchi adeeza would not light because the two structures were both male in nature. It was because of that imbalance that the fire would not light, and so a female version of the hogan had to be constructed. Since that time, both male and female hogans have been made. There is no real significance in which style the Navajo choose to build. From my observations, I would say that the female version of the hogan is predominant, yet the alchi adeeza is greatly respected. It is a reminder of how the Holy People once ascribed this kind of home for us, and also a reminder of our ancestors, who roamed the southwestern landscapes so that future generations would inhabit those same lands.

In Navajo tradition, fire is considered sacred because it provides heat for human comfort and cooks the food that nourishes us. As Diné ancestors learned, fire, too, was classified into four different types, and each had its respective gender. That fact is not generally known among the Navajo people. This kind of knowledge can be gleaned only from listening to songs, prayers, and recitations of the Holy People's history during the traditional Navajo healing ceremony.

Navajo hogans are designated male and female because everything in the natural world has a male and female aspect. It is through this that nature and human beings regenerate. Navajos assign paramount importance to natural balance, and they achieve this by according female and male aspects to all organic material in the natural world. The round hogan is designated female and the fork stick style is considered male. Both hogans can be used for habitation and ceremonial purposes, and both can be memorialized as tombs if someone passes on and is buried in a hogan.

Another crucial factor in Navajo thought is the importance of the four directions. Diné believe that they were given lands within four sacred mountains, along with four sacred rivers that bound this land. These sacred mountains are Sisnaajini(Sierra Blanca Peak) in the east, Tsoodzil (Mt. Taylor) in the south, Dook'oosliid (San Francisco Peak) in the west, and Dibe' Ntsaa (La Plata Mountain) in the north. Each of these sacred mountains is assigned a gender, a precious gem, a color, and plants and animals. Each thing associated with the sacred mountains is native to its specific area. The respective gems, soil, plants, and animals of these sacred mountains are integral to the highly sacred healing bundles that are property of Diné traditional healing chanters from the eighty or so different clans.

Diné believe that the mountains serve as markers to their lands within the four sacred mountains. This homeland is called Mother Earth. They believe that living according to the natural laws of Mother Earth ensures that life continues to be balanced and that the state of *hozho* (harmony) is achieved. The state of hozho implies that there is an abundance of necessities for children and for their children's children, and everything will be plentiful for the future.

The sun is considered the Father in the natural world. It is Father Sun who brings light into the daily lives of human beings, other living creatures, and the plant world. The laws of the earth, sun, and moon, plus the laws of organic living matter, prescribe the order and harmony that are paramount in the Navajo belief system.

The need for order can be seen if one understands the importance of the four sacred directions in the Navajo world. The four directions of Diné bike'yah (Navajolands) can be transfigured into the human lives and order in the hogan. The east signifies all that is good in the universe. Hence the entrance always faces east. Navajo families desire that the first light enter through the entrance because humans beings, by natural law, are supposed to function in light. When the first light enters the hogan, the thought processes of human beings begin clicking. It is believed that the beneficent holy deities travel on the earliest rays of the sun, and Navajo people teach their children to wake up before the sun. A practice of awakening at dawn ensures that the deities will be familiar with and recognize Navajo children. In this way, the deities look favorably upon the children who will soak up the beneficence of everything that is fresh, new, and good.

The south side is like a storehouse of knowledge. It represents the early life stage of an individual. This side is also considered beneficent in the rearing of children from birth up until the time that they are considered mature and responsible.

The west side represents the adult phase of an individual. This is where a family's goods are kept. The west side of the inside roof of the hogan is the place for a medicine bundle, which includes the *dzil leezh* (sacred mountain soil). This is kept on the west side of the hogan because a traditional healing chanter wants most of the light of the day to bring warmth and beneficence to the sacred bundle. Continuous bathing of the light ensures health to the ceremonial bundle, which in turn blesses the chanter, the family who act as stewards, and potential patients.

The north side is the prescribed direction for malevolence or evil. This is evident in Navajo healing ceremonies, in which a traditional chanter has the power and the knowledge to pray over one who is being affected by malevolent energy. A traditional chanter will pray and sing and "blow the evil away" from his patient. The evil is blown to the north. The north is also designated for the later stages of a person's life. It is considered a time when a person has experienced the good and the bad in life. This section signifies the aged. It is a time for retirement, leisure, and slow living. During normal habitation of a hogan by a Navajo family, hunting tools and other sharp objects are stored on the north side of the hogan.

As one can see, the circularity of the hogan is important because each stage is represented by one of the four directions. Life is sacred to Navajo people, and order is set by the laws of nature. The sun comes up from the east and sets in the west. The path of the sun is considered the primary natural law; therefore, during sacred times like healing ceremonies, eclipses, and crescent moons, it is customary for human traffic to mimic the direction of the sun. This means that a person enters at the entrance and goes to the left, sitting on the south side of the hogan if a man or on the north side if a woman. When exiting, the person exits again to the left. By doing this, an individual completes the circle. This ritual is very important during traditional healing ceremonies. Besides following ceremonial protocol, this practice eliminates the collision of human traffic.

Navajos adhere to the natural laws even in the construction of hogans.

They build hogans so that the logs are set in the upwardly growing position. This can be seen in the construction of the fork-stick style as well as the female many-legged hogan. In the construction of male hogans, logs are set upright, again in congruence with the natural upward growth of trees. The law of the sun's path plays an important role in the construction of the female hogan, which employs cedar logs. The horizontal layering of logs is done in such a way that the logs mimic the sun's path.

The Diné were gifted with the two styles of hogan by the holy deities, and earthly people believe that the deities act in their behalf for beneficence. The Diné people were instructed to live in these structures for time immemorial. The roundness of a hogan can be construed to resemble the vast horizons of the Southwest. When one is out in the open lands, one can see for miles, and it feels as if one were encircled by a vast sky. A better description might be that one feels covered by a blue bowl. The predominant feature of the southwestern Diné homeland is its vast landscape with open blue skies. This may be one reason why the landscape and the cosmology have a significant role in the oral traditions and sacred ceremonial text of the Diné. In addition, there is tremendous importance placed in the directions, the natural laws, and their role in the orderliness of Navajo society and culture.

The Navajo people are very practical, and their lifestyles reflect that fact. This can be seen in the construction of a hogan. A hogan faces the east so that the sun will illuminate the inside at first light, awakening the inhabitants so that they may go about their daily routines. The hogan is also situated so that it deflects the prevailing southwesterly winds. Hogans are practical because they are constructed of native materials such as cedar logs, stones, bark, and mud. Through the use of these materials, hogans are organic structures that offer an attractive option for natural living.

In the old times, hogans were supposed to be constructed by a new son-in-law within four days after he married into his wife's family. Upon building, he was supposed to plant *nitliz*, precious stones that are associated with the four directions. The stones were placed in the ground under four of the main posts in relation and in respect to the four directions. When the new in-law had constructed the hogan, he was supposed to have a Blessingway ceremony first and then allow his wife to move in and start the first fire. It was significant that she light the first fire because the new hogan was considered her hearth for the children who would come from the union. The confines of the hogan have always been considered the woman's domain. She has the responsibility for providing meals and warmth and maintaining orderliness so that her children will see, experience, and practice the teachings of the culture. The wife will instruct her children to observe, learn, and act according to the guidance and example of the adults. The children will be taught to obey and to respect the teachings of Navajo culture.

It is said that men are keepers of the entrance to a hogan and beyond. This means that they are responsible for all other work that takes place outdoors. Some of these responsibilities include hunting, farming, animal husbandry, and more contemporary wage work. Men, along with women, have to teach children about the home, the outdoors, and the environment.

Hogans are energy efficient because the circle is the most efficient use of space. The circular shape of the hogan ensures equal distribution of light and warmth, which are at the center. The circular room is warmed such that inhabitants get the most use of the heat necessary for surviving the harsh winters. In the summer, the thick roof of the hogan's terra cotta–like covering and the thick insulative walls keep the heat out and maintain a cool, comfortable temperature inside.

Hogan construction costs can be minimized by using materials from the nearby woods. However, with the advent of contemporary construction techniques and the onslaught of federal housing regulations through the Department of Housing and Urban Development (HUD), Navajo people have unfortunately been moving away from the native traditional values pertaining to hogans and, more important, away from living in harmony and in balance with Mother Earth. The Diné are gradually abandoning the traditional and efficient living style that requires only minimal necessities.

Today, Navajo people are burdened with having to purchase furniture and many other items that supposedly make modern life "easy." Many contemporary Navajo people are living like all others in the dominant society. The change in the shape of their homes has caused rifts in traditional family dynamics. Children no longer live within the same space as their parents or their grandparents. Many elders concerned about the growing problems stemming from changing lifestyles are troubled about the degradation of traditional Navajo values and beliefs.

Hogans served as homes for human beings as well as beneficent spirits. For Navajo families, hogans are considered extensions of Mother Earth. Some make analogies that their hogan is like the womb of a mother. It is within this womb of Mother Earth that Navajo women create a nurturing environment for the upbringing of children. The roundness of the hogan is likened to the protruding tummy of an expectant woman. The circularity of the hogan is congruous with the soft lines of the woman. The parallel logs are like intertwined fingers protecting and encircling a nurturing space within the universe.

The *chi ladei* is an opening at the top of the hogan. Through it, smoke from the fireplace passes into the air. The chi ladei can be likened to the umbilical cord that provides nourishment for a baby. Similarly, the chi ladei works in tandem with the entrance to circulate fresh air. The opening in the hogan also allows those inside to see the changing star patterns in the night sky, which can be read to tell the time or season.

Traditional hogans do not have windows. However, it is not unusual to encounter a structure with one or more windows. Spiritually, the malevolent energy is concentrated in the north wind, which hits the north side of a hogan. When a Navajo passes on, a hole is knocked in the north wall to let the malevolent energy out. A hogan with part of the north wall knocked out indicates that a death has occurred and that people should not enter, out of respect for the deceased and the family. Such a hogan is no longer considered a living place, and it should be avoided.

Hogans can also be religious structures for ceremonial healing purposes. A chanter will come to someone's home and perform healing ceremonies on a patient for a certain number of days. During the ceremony, the chanter sings, prays, and performs other sacred activities such as sandpainting. When a sandpainting is created, it is very personal and sacred. The making of a sandpainting on the floor of a hogan sanctifies it. The hogan is unique and sacred from that time on. At a prescribed time, the sand used in a sandpainting will be removed and deposited at a specially designated place where desecration will be only a remote possibility. When a hogan is being used for healing the sick, it is a sacred place, and special protocol should be observed by everyone.

The history of the hogan is rooted somewhere in the Navajo past. I am not aware of the various stages of its development, but I know that there is research material on the subject. Today the hogan continues to serve as shelter for Navajo families across the Navajo reservation in Arizona, New Mexico, and Utah. Hogans continue to meet the needs of my people. It has taken thousands of years to perfect this structure, and it continues to undergo changes to reflect modern times. Navajos enhance the traditional style of the hogan for aesthetic reasons as well as to enjoy modern conveniences. This is evident in a variety of hogans that have two stories, or wings that extend off the sides to create additional spaces for bedrooms and kitchens. And the idea of the hogan is becoming more accepted beyond the reservation. Both Navajos on the reservation and non-Navajos have built and live in this traditional home.

The concept of the hogan as a traditional home is broadening to allow for additional features that will enhance its appeal. The hogan presents a flexible idea: The only stricture is that it sit so that its entrance faces the east. Navajo people continue to experiment with designs and employ new and widely available materials in their hogans. I believe that as long as the Navajo people rise before the sun, pray to the deities, and perform their sacred ceremonies, they will continue to make use of the hogan as prescribed by the holy deities. By living in and using the hogan, Navajo people are reminded of their ties to Mother Earth and their duty to ensure that there is balance and harmony within themselves, their environment, and the universe.

As a symbol of beliefs and values for Navajo society and in its importance to the way we live with our environment, the Navajo hogan was devised as a blueprint for the continual succession of Navajo people from one generation to another. These simple, mud-plastered structures represent a sophisticated plan for a people who believe that harmonious existence with the world is the first law of the universe. Living with orderliness ensures harmony. When Navajos live according to the natural laws, they in essence are "walking in beauty," which is a predominant theme in prayers and songs. By "walking in beauty" they achieve *hozho*, a state of being in which everything is harmonious, peaceful, and happy.

4

The Agronomy of Memory and the Memory of Agronomy

Ritual Conservation of Archaic Cultigens in Contemporary Farming Systems

MICHAEL R. DOVE

Throughout upland southeast Asia, plants are cultivated and revered that were once economically important but that have now been superseded by other crops, making them economically marginal. Jacques Barrau, in a 1965 article in the journal *Ethnology*, asks regarding these plants, "Why, then [are they] still used and occasionally cultivated?" (Barrau 1965b:288). To put this question more prosaically, Why do people plant and tend archaic cultigens when they have no obvious economic need to do so? Barrau cannot answer this question, but he may unknowingly point in the right direction when he refers to these plants as "witnesses of the past."

We might rephrase Barrau's question as, Why is it important to preserve within a field some element of the crops and farming systems that gave rise to it and say something about its history? Relatively little attention has been paid to this question, perhaps for the same reasons that relatively little attention has been paid to indigenous history of any sort.[1] As Barrau (1965b:282) observed for Oceania (in another prescient statement): "There was a tendency to assume that food-plant patterns found by Europeans . . . were representative of subsistence economies established in ancient times. Attention seldom was paid to the changes which may have occurred in the economic or cultivated flora of pre-European Oceania." Like other histories, the histories of indigenous crops were denied. And when these histories were not denied outright, they were viewed selectively. Thus, Condominas [1972:56] writes, "Western ethnocentrism has resulted in the origin of agriculture being seen in the domestication of cereals," ignoring the earlier history of domestication of tubers.

Indirect, physical evidence of this history is increasingly being studied by archaeologists; but some of the most intriguing evidence, as suggested by the opening citation from Barrau, is that which is intentionally preserved by contemporary peoples in the form of archaic plants and the myths, rituals, and language pertaining to them. Whereas scholarly attention to these matters has not been wanting, systematic scholarly attention to these data, for the express purpose of "reading back" into the agricultural prehistory, has been surprisingly limited—with a few notable exceptions. Thus, Conklin (1959: 300) long ago urged greater attention to the vernacular plant names of the Pacific/Indo/Malaysian region, for the indirect evidence they could provide of the "pre-European spread of cultigens, weeds, and agricultural practices." Scaglion and Soto (1994) illustrate the potential of such studies in their use of data on the vernacular names for sweet potatoes to contest the theory that this crop was introduced into New Guinea from the West and during the European period. Fox (1991) uses data both on plant names and the ritual contexts in which they are employed to make some interesting inferences about agricultural prehistory in Eastern Indonesia.

The idea that plant-use in contemporary ritual contexts may say something about agricultural history is especially intriguing, and it has suggested itself to a number of observers (Condominas 1986:44; Harris 1977:215; Pelzer 1945; Spencer 1966),[2] although it has infrequently been pursued in depth. An example of how such a study might develop is given in Coursey's and Coursey's (1971:478) analysis of the "New Yam festivals" of West Africa (cited in Harris 1973), one of the conclusions to which is that the proscription on iron tools during these festivals points to a pre-iron age origin for yam agriculture.

The "mapping" of parallel developments in culture and agriculture, by anthropologists and others, is not new; but to date, most such efforts have been biased in favor of illuminating the human history as opposed to the plant history. These efforts are exemplified by Rival's (1993) and Sugishima's (1994) stimulating analyses of botanical metaphors to decipher human descent and alliance. My intention here is not simply to urge a reversal of emphasis. I suggest that (1) it is both possible and productive to take a less essentialist approach; (2) archaic plants tell us not just about the history of current crops but also that this history is important to their cultivators; (3) this cultivation is a way of mediating between past and present, an important mechanism of cultural memory and history; (4) these plants are not just "witnesses of the past" (as Barrau suggested) but also witnesses for the present; (5) what is unique and important about these archaic plants is the way that they exist in and yet transcend the present and its concerns; and (6) this transcendence is relevant to the contemporary global concern with conservation of biodiversity and promotion of sustainable resource use.

I will begin my analysis, after first presenting some background data on present and past agriculture in the region, in particular on the island of Borneo, by discussing the contemporary role of taro (*Colocasia esculenta* [L.] Schott) as a minor starch staple, and the cultural evidence of its role as one of the earliest domesticates. I will then discuss two grains, Job's tears (*Coix lachryma-jobi* L.) and Italian millet (*Setaria italica* [L.] Beauv.), which retain largely ritual importance today, and I will argue that this ritual character provides evidence of their precedence to rice (*Oryza sativa* L.). Finally, I will suggest that in all of these cases agricultural succession is represented as a matter of historical contest, not of "disappearance." This framing of agricultural development has transcendental qualities, thus overcoming the obstacle that social determination poses to thinking about social problems, including those of global environmental thinking and management.

Contemporary, Historical, and Prehistoric Agricultural Patterns

Most of the data for this study come from Borneo, with supplementary data on other parts of the region, including Java and the Malay Peninsula. Within Borneo, the focus is on the Dayak, and on one tribal grouping in particular, the Ibanic-speaking Kantu' of West Kalimantan (fig. 4.1). Most Dayak communities have today, and have had for centuries, "composite economies." These economies consist of swidden-based production for subsistence purposes, coupled with swidden- or garden-based production or forest product–gathering for market. The primary subsistence crops are dry rice (as well as some swamp rice) and a wide variety of nonrice cultigens, including a second cereal, maize, a number of different cucurbits, and a number of tubers, including cassava and sweet potato and, to a lesser extent, taro. Rice is the primary starch staple and the focus of each meal, supplemented with one or more of the nonrice cultigens as a "relish."[3] When the rice crop fails to meet subsistence requirements (as it may several years in ten), the market crops are sold to buy rice; if that fails, then cassava and other tubers become the famine-period starch staple. Subsistence needs aside, a number of additional plants, includ-

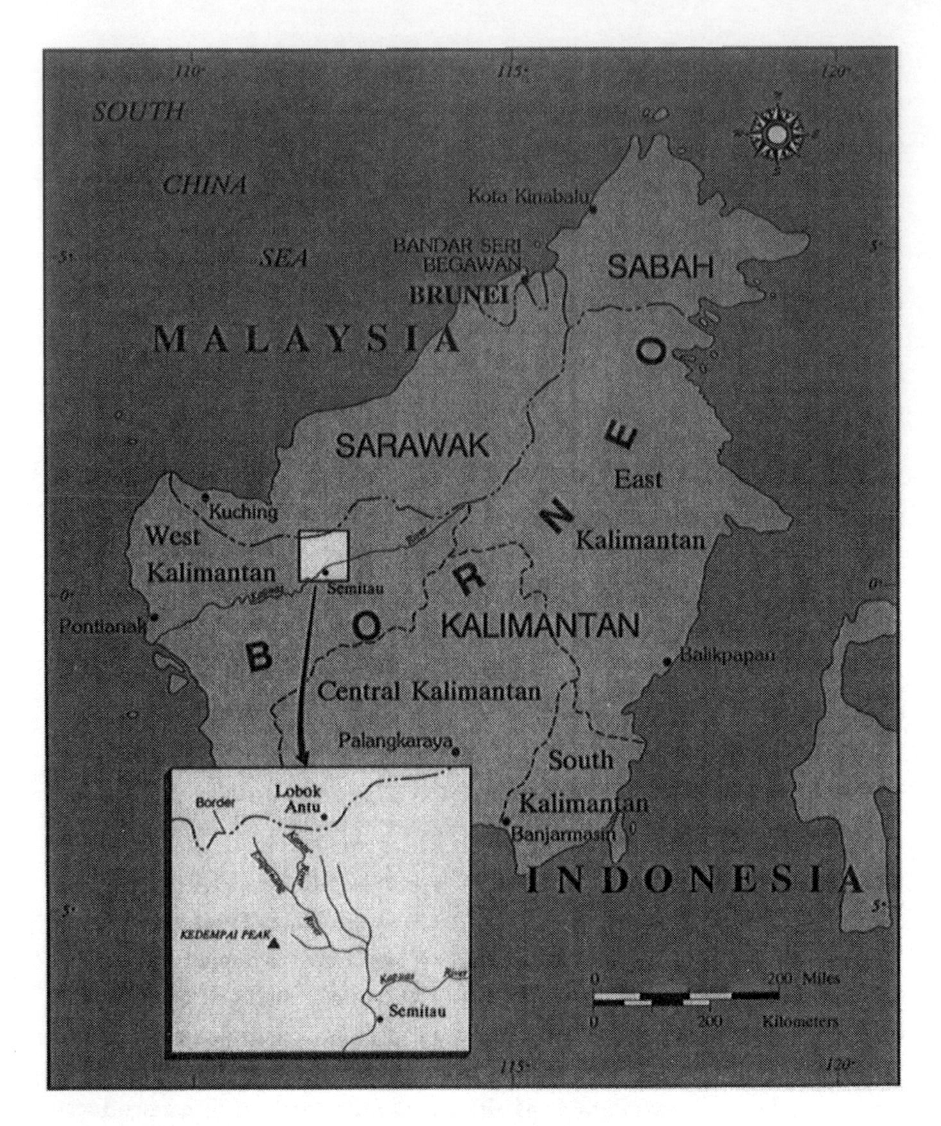

Figure 4.1. The island of Borneo and the Kantu' territory.

ing Job's tears and Italian millet, are cultivated in small amounts primarily for ritual purposes.[4]

This still-ubiquitous upland-farming pattern is but the latest "moment" in a long sequence of changes. These changes have not been confined to market- oriented activities, as the scholarly focus on the dynamic and ancient trade of the region might lead us to expect. On the contrary, some of the most important changes have taken place within the subsistence sector, which we tend to think of as more static. Even a review of Dayak names for rice varieties, which frequently allude to their origins in distant lands (Dove 1985b:161), tells us that the subsistence agricultural sector has been the subject of a great deal of diffusion, adoption, and innovation. The dynamic nature of this sector should also be evident from the oft-overlooked fact that many of the staple crops of interior tribal people like the Dayak—such as maize and sweet potato, in addition to the market crop rubber (*Hevea brasiliensis* [Willd. ex A. L. Juss.])—are introductions from the New World. Some of these introductions (such as sweet potato and maize among the Kantu' Dayak) have been so well integrated into the indigenous system of cultivation that they have acquired their own origin myths, being ascribed to the *antu* ("spirits").[5]

Although some of the plant introductions just discussed may have taken place centuries ago, they represent relatively recent changes compared with some others that have occurred. There is evidence, for example, that in the history of the dryland cultivation of cereals in many parts of southeast Asia, rice was preceded by other grains, namely Job's tears and various millets (Harlan 1992:206–7).[6] The antiquity of millet cultivation in southeast and east Asia is manifested in the fact that the ancient names for at least two countries, Java and northern China, mean "land or island of millet" (Li 1970:8; Raffles 1978, 2:67; but cf. Strickland 1986:165).[7] There also is evidence that dry-land cereal cultivation itself was preceded in some parts of the region by wetland cultivation of tubers, in particular taro, which is perhaps the oldest cultigen in southeast Asia (Barrau 1965a; Burkill 1951; Harlan 1992:207). Part of the historical evidence for this is the absence of rice in Oceania, which suggests that rice cultivation was not part of the technology of the early Austronesian-speaking peoples who migrated to, and populated, Oceania about five thousand years ago (Bellwood 1979:141; Pelzer 1945:6–7).

The available evidence suggests that there have been two major dimensions of agricultural evolution in southeast Asia (Barrau 1965a; Sauer 1952; Yen 1971). One involves wetland versus dry-land agriculture and the other, overlapping the first, involves vegetatively reproduced crops in perennial gardens versus the seed-based reproduction of crops in shifting cultivation fields (Chesnov 1973:5; Harris 1973:398; fig. 4.2). While wetland, vegetative production may have preceded dry-land, seed-based cultivation in the prehistory of some parts of the region, the two represent not so much successive stages of development as two separate (albeit interdependent) developmental tangents. Thus, there were and are societies—like the Kantu' Dayak—that have combined elements of both traditions in their systems of agriculture.[8]

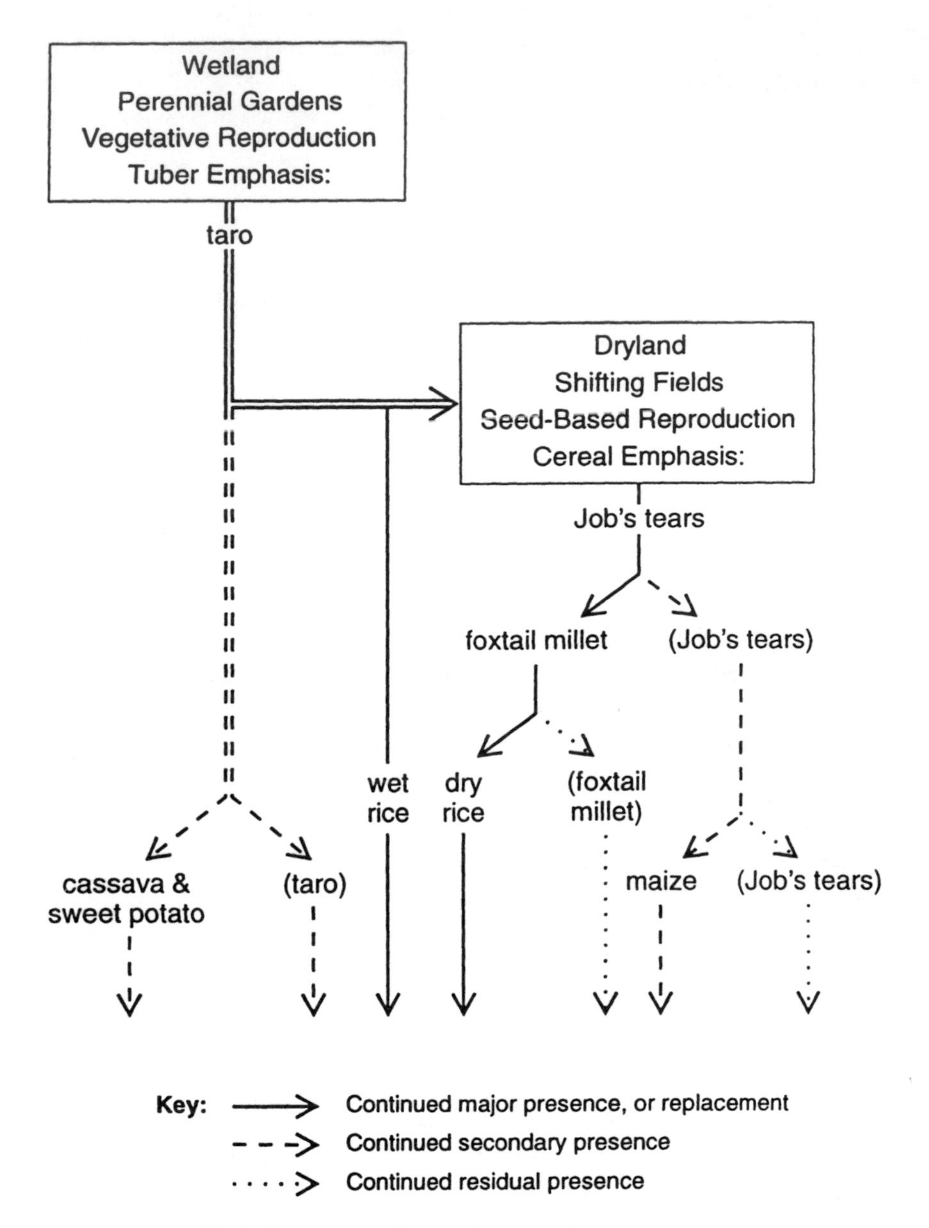

Figure 4.2. Two likely dimensions of agricultural evolution in southeast Asia.

The Tuber-Grain Transition

The historical transition between tuber cultivation and grain cultivation is reflected in the oft-observed discontinuity between the economic role of tubers in contemporary cultivation and their role in ritual. This discontinuity, in turn, is the product of the common tendency in the region for newly adopted cultigens to push the cultigens that they replace into ritual niches. These ritual niches are extremely tenacious in the face of time and change—perhaps because it is time and change that create and define them.[9]

The Contemporary Status of Taro

One of the most important of the early tubers of southeast Asia, taro, is still grown throughout southeast Asia, although it is not as widespread as it apparently was in prehistory. Spencer's mapping (1966) of the distribution of taro as a staple crop shows that it extended as far west as India in prehistory, that it had retreated to the east of Borneo by A.D. 1500, and that it was no longer found (as a staple) west of New Guinea by A.D. 1950.

The contemporary Kantu' of West Kalimantan, for example, cultivate taro as a minor "relish" in swamp swiddens and in swampy, low-lying parts of dry-land swiddens. It is less important to them as a food crop than either of their grain crops, rice and maize, or either of the other major root crops, sweet potato and cassava.[10] Taro is distinguished in this sort of agricultural system in part for its great diversity at the varietal level—the Kantu' recognize seven different named and still-cultivated varieties of taro, which is exceeded only by rice (forty-four varieties), bananas (twelve varieties), and sugar cane (ten varieties), but exceeds, for example, cassava (four varieties) and sweet potato (one variety)—and in part for its prominent role in mythology concerning the origins of culture and agriculture.

Mythological Evidence of Early Taro Cultivation

Taro is prominently mentioned in Dayak mythology, and much can be inferred about prehistoric taro cultivation from an analysis of this material. Taro is associated in myth with a major deity of the Kantu' and other Ibanic-speaking Dayak, namely Pulang Gana, believed to be the first human being who ever lived. According to traditional belief, Pulang Gana is the *Kuasa tanah*, "guardian of the earth." According to mythology, Pulang Gana originally subsisted by eating charcoal, like his nonhuman spirit ancestors, and then he *nemu keladi* ("found taro") and subsisted by eating that.[11] It is only later in most versions of this mythological genealogy (three generations, in one popular version) that rice is first grown and eaten, and Pulang Gana becomes the *Rajah padi* ("king of the rice").

The myth of Pulang Gana is a history (and a clear cultural statement about the relative antiquity) of taro cultivation by the Dayak. The precedence of taro over the other tubers, cassava and sweet potato, is reflected in the fact that the origin of these other tubers is merely attributed to the *antu* ("spirits") in general; whereas taro—alone among cultigens—is attributed to Pulang Gana, the first ancestor. And the precedence of taro to rice is clearly reflected in the narrative structure of the myth itself.[12] (Note that the acquisition of rice cultivation is presented as following the acquisition of taro but not necessarily as replacing taro, and that this myth says nothing about abandoning or getting rid of taro.)

The fact that the first food of the mythical ancestors of the Dayak is believed to have been charcoal may also be relevant to the historical precedence of tuber cultivation to grain cultivation. The physical difference between a meal composed of numerous small grains and one composed of a few large tubers has implications for the cooking technologies that can be employed. One of the most efficient (and also digestibility-enhancing [Stahl 1989:181]) methods for cooking numerous small grains is to boil or steam them in a water-filled container, which is the method employed by the contemporary Dayak for almost all consumption of rice.[13] This is not equally true for tubers, which can just as easily be roasted in direct contact with the fire—as is still done today among the Kantu' for example, with perhaps one-half of the cassava that is eaten.[14]

In historical and prehistoric times, before metal cooking utensils reached the interior tribes of Borneo, the practice of roasting tubers directly in the fire is likely to have been even more common than it is today.[15] And it is possible that the period in their history when the Dayak subsisted on fire-blackened tubers has come to be remembered in myth as the period when they subsisted on charcoal.[16] It is reasonable to expect the developmental gap between eating charcoal and eating boiled grains to be culturally marked in some manner: The elaboration of techniques for processing plant foods, although overlooked by many scholars, can rival the elaboration of agricultural techniques in its contribution to the intensification of subsistence systems (Stahl 1989:185; Yen 1975:162). (In some contemporary [or recent historical] ritual contexts, the Dayak mark this developmental gap by collapsing it: They boil rice in a mixture of soot and water [Roth 1980, 1:415–16].)[17] If this interpretation is correct, the mythical reference to charcoal-eating also attests to the precedence of tuber cultivation.

The historical precedence of tuber cultivation is also implied in the generational connotations of some contemporary crop metaphors. The parent-child relationship is a common idiom in the region for crop succession. Thus, among the Semai of the Malay Peninsula, Job's tears is called the "mother" of the Italian millet (Dentan 1968:47; cf. Condominas 1972:53). A similar relationship is implied in Dayak mythology about the origins of rice. According to Kantu' oral mythology, the first human being to eat *nasi* ("boiled rice," or "boiled grain") was Bui Nasi. He did not want to eat charcoal, which up until then had been the staple food of humans (with the later addition of taro), so he asked his parents for grain. They replied that they would have to die to give him grain, to which he replied that he still wanted it.[18] So his parents assented, informed Bui Nasi that rice would grow from their grave, instructed him on how to cultivate it, and then died. After three days, two panicles of rice appeared on their grave; Bui Nasi weeded and fenced them, and that was the beginning of cereal cultivation.

The term used for "grave" in this story is not the expected Kantu' term *pendam* but the Malayic term *kubur*. In its active form, *kubur* translates not only as "to bury" (Scott 1956:94) but also as "to heap up soil, ridge (as sweet potatoes), build up vegetable or seed bed . . . for drainage" (Richards 1981:167–68). This association between human burial and techniques of tuber cultivation also is suggested by a contemporary Iban prescription for filling graves so that they look like raised mounds, or, in effect, like root-crop mounds (Richards 1981:268).[19] In this context, Bui Nasi's burial of his parents to obtain grain can be likened to the burial of a tuberous plant for vegetative propagation; and this tuber mound can be seen as responsible for the subsequent growth of the first grain plants. This complex of associations suggests that parents are to children, therefore, as tubers are to grains—or as tuber cultivation is to grain cultivation. The generational dimension and tension of the Bui Nasi story may also be reflected in the use, among young Dayak today, of the phrase "*tuai ubi, tuai keladi*" ("old tuber/cassava, old taro") as an expression of disrespect for elders.[20]

The use of tubers to symbolize "dim-wittedness" is in accord with the fact that less knowledge (and less cultivation, and less culture) is required for the vegetative-based reproduction of tubers than for the seed-based reproduction of grains. For example, one of the most critical pieces of knowledge in contemporary rice cultivation is the proper date for planting (and therefore harvesting) the rice. The proper dates are relatively narrow in range, and too much error in either direction can cause the rice crop to fail. The dates are calculated on the basis of a variety of esoteric phenomena, which include the breeding cycles of certain animals, the phases of the moon, and the movement of certain constellations in the night sky. In contrast, in root-crop cultivation, neither the date of planting nor the date of the harvest is critical: They can be varied at will without ill effect. On this basis, at least, root-crop cultivation requires less knowledge than does grain-crop cultivation.[21]

If root-crop cultivation is more "ignorant" than grain-crop cultivation, the latter is perhaps more "greedy" than the former. Annual grain crops generally place more demands on planting environments (Harris 1972:188), in return for lower yields than most root crops.[22] As a result, land that is cropped in grains usually must be cropped for shorter periods and fallowed for longer periods, and more of it must be cultivated, than land that is cropped in tubers. In taking so much more out of the land and in yielding so much less, grain production is thus in fact "greedier" than tuber production.[23] This characteristic seems to be reflected in the myth of Bui Nasi in the fact that his desire for rice comes at the cost of his parents' lives, and in the very name of the child who insists that this price be paid, Bui Nasi, which means "evil spirit of the rice."

There is a further explanation for these views of tuber versus grain cultivation, however, that has less to do with the demands that are placed on soils and cultivators by their crops than with the demands that are placed on soils and cultivators by the wider society. Many of the early states in the region were based on agriculture, and the exigencies of state formation and state support have played a major role in agricultural development. Of particular importance is ease of control and extraction, which is reflected in the support given by states and state ideologies to cropping systems that facilitate, as opposed to inhibit, these relations. This factor can be seen in the exaggerated support versus condemnation that characterizes the attitudes of southeast Asian states to this day toward the two predominant systems of grain cultivation in the region: pond-field versus swidden cultivation of rice, respectively (Dove 1985a).

There is an analogous distinction with respect to the cultivation of grain versus tubers: Just as rice swiddens are less susceptible to extraction than rice pond-fields—because the former are less concentrated and permanent than the latter—so are tubers generally less susceptible to extraction than grains—because they are less easily stored and transported. This distinction in susceptibility to extraction has tended to be represented in the overarching state ideologies of the region as a distinction between the low culture (of tubers) and the high culture (of grains), a distinction that has perhaps been internalized even by groups like the Kantu'—as expressed in their cultural focus on rice and their disparagement of tubers—that have traditionally lived at the margins of the state. On the other hand, the fact that the Kantu' also portray rice as greedy suggests that such groups also possess some critical awareness of the downsides of grain cultivation.

State interest in extraction was not the sole driving force in the succession of tubers to grains, however: Another factor is nutrition. A number of scholars (such as Harris 1972:188) have noted that tubers tend to be less nutritious (in particular with respect to proteins) than cereals, and as a result hunting was a more necessary accompaniment to tuber cultivation than to grain cultivation. It is noteworthy, therefore, that hunting is often associated with the tuber-grain transition in indigenous mythology. Among the Kantu', for example, the story of the first acquisition of rice culture is told, in a variant of a tale told all over southeast Asia, as follows:

Pulang Gana's taro was suffering the predations of a porcupine. So he set a spear trap near his taro. The next morning he saw that the trap had been sprung. Seeing blood from the porcupine, he then set to tracking it with three of his dogs. Pulang Gana tracked the porcupine until the following morning, without success, when he came upon the longhouse of Rajah Swa. Eventually Rajah Swa gave his daughter in marriage to Pulang Gana and he placed Pulang Gana in control of the success or failure of all rice swiddens: Thereafter, no farmer's swidden would succeed unless Pulang Gana was first informed [by means of ritual offerings] of the farmer's intentions.

There are two notable points in this myth. First, hunting is specifically associated with the era of tuber cultivation, which is set prior to the era of grain cultivation—an idea in accord with the scholarly speculation that hunting was more of a (nutritional) necessity with tuber than with grain cultivation (Harris 1972:188; 1973:403). Second, it is the act of hunting that leads to the acquisition of grain cultivation. That is, it is the search for—and hence need of—the proteins and other nutrients missing in the tuber-based diet that leads to the succession from grain cultivation to tuber cultivation. The succession is thus explained as a response to the protein deficiencies of tuber cultivation, which are made up through cultivation of the more nutritious grains.[24]

The Transition from Archaic to Modern Grains

Just as there is a discontinuity between the role of tubers in contemporary agriculture and their role in ritual, so is there a similar but even more marked discontinuity for the grains. Indeed, it was this latter discontinuity that first drew my attention to the agricultural prehistory of upland southeast Asia.

The Contemporary Status of the Nonrice Grains

There is evidence, as noted earlier, that rice was not the first grain to be cultivated in many parts of southeast Asia—that it was preceded by other grains, in particular by Job's tears and Italian millet. There also is evidence that Job's tears is more archaic than millet (as suggested by the reference in the previous section to the Semai belief that Job's tears is the "mother" of Italian millet).[25] It appears likely in many regions that rice has succeeded millet (Hutterer 1983: 183n.10; cf. Skeat and Blagden 1960, 1:11, 341, 343), and millet itself had earlier succeeded Job's tears (Condominas 1972). (Many observers also suggest that the arrival of the New World grain, maize, provided a further substitute for Job's tears [Arora 1977:365; Burkill 1966, 1:629–30; Condominas 1972:48].)[26] Job's tears is an appropriate candidate for one of the first grains to be domesticated and cultivated: It is rich in carbohydrates and starch (Burkill 1966, 1:638–39; cf. Hutterer 1988:69)—indeed, it is more nutritious than rice (Crevost and Lemarie 1901–1902:30 [cited in Condominas 1972:50])—and it grows wild in disturbed areas of the southeast Asian forest (Dentan 1991:426–27).[27] It is also a true multipurpose plant, having value as food, medicine, magic, fodder, and ornamentation (Condominas 1972:50).[28]

The former importance of these two grains in southeast Asia is much diminished today (although little is actually known about how this came about).[29] There are pockets of millet cultivation, for use either as food or bird seed;[30] the cultivation of Job's tears, largely for ornamental use, is perhaps even rarer.[31] The Kantu' grow very small amounts of both grains in their upland swiddens, but not as staples or relishes (unlike taro). They occasionally use Job's tears to brew liquor (Arora 1977:362–64; Sather 1977:167n.13; Watt 1966:396), and they sometimes use Italian millet to make a special porridge to serve during feasts. But the most important role of these two archaic grains in Kantu' society is in ritual;[32] it is here that the antiquity and former importance of these grains is addressed.

Ritual Evidence of the Transition from Archaic Grains to Rice

There are two times during the annual swidden cycle of the Kantu' when exceptional status—completely incommensurate with economic importance—is accorded the two archaic, nonrice grains, Job's tears and Italian millet: The first is during planting, and the second is during the harvest and postharvest ceremonies.

There is widespread belief in southeast Asia about the potential for "conflict" between rice and the more archaic grains.[33] Among the Kantu', this belief was formerly manifested as a proscription against planting Job's tears and Italian millet too close together, for fear that their "conversation" would "frighten" the spirit of the rice. (Sather [1977:167] reports a similar, extant belief among the Saribas Iban, who apparently plant larger amounts of Job's tears than the Kantu'. It involves a proscription against harvesting any Job's tears before the ritualized first harvest of the rice.) I suggest that this belief in the ability of the two archaic grains to frighten the rice relates to the postulated historical replacement of rice as the grain starch staple by Job's tears and Italian millet.[34] This history explains why the "spirit" of the rice is perceived as anxious about what the archaic grains might be saying (or plotting) about it.[35]

The fact that separation between Job's tears and Italian millet is also an issue in this planting proscription may reflect the likelihood that the former preceded and was replaced by the latter in many agricultural systems in the region. That the proscription was once deemed necessary at all suggests, first, that there was a time when Job's tears and Italian millet were both cultivated in significant amounts, in the same fields (namely, during a period of transition from the former to the latter); and second, that there was also a time when these two grains were cultivated in significant amounts and in the same fields with rice (namely, during a period in which rice was gradually supplanting the nonrice grains).[36] If neither of these scenarios had ever occurred, there would have been no need for the planting proscription.

That similar planting proscriptions do not apply to other crops also is revealing. For example, there are no such proscriptions involving taro, which probably reflects the fact that taro is still grown as a food crop by the Dayak (although in diminished extent, redefined from a starch staple to a relish).[37] The lack of proscriptions involving taro also may reflect the fact that the tradition of tuber cultivation in general persists, although again in diminished extent, alongside the ascendant grain cultivation tradition. Of equal interest is the absence of any planting proscriptions involving maize. The presence in

Indonesia of societies in which maize has become the grain starch staple (such as in montane parts of Java and the islands of eastern Indonesia) demonstrates that maize can threaten the place of other grains and other cultigens (Fox 1991). The fact that the Dayak have not surrounded maize with ritual proscriptions may thus suggest that these proscriptions are not normative statements about potential agronomic conflict; rather, they are empirical statements about actual historical competition.

Each year's swidden harvests among the Kantu' begin with the ritual first harvest of the *padi taun*, or "annual grains." This involves harvesting a handful of panicles from each of four crop types: *padi* ("dry rice"), *puloi* ("glutinous rice"), *lingkau lesit* ("Job's tears"), and *jawa'* ("Italian millet"). In the use of the term *padi taun* for these four grains, *padi* clearly denotes grain and not just rice; and the qualifier *taun* ("year") seems to denote annual crop as opposed to perennial tree or tuber crop.

It is noteworthy that this grouping excludes *lingkau keribang* ("maize"), since that is also an annual grain, and shares sufficient other morphological characteristics with Job's tears to share a lexical root as well (cf. note 26). I suggest that this exclusion is a statement about the relatively recent addition of maize to Kantu' agriculture. It is equally noteworthy that the *padi taun* include Job's tears and Italian millet with the glutinous and non-glutinous rices, given that the latter are the mainstay of the Kantu' diet, while the former have virtually no dietary or other direct economic importance. I suggest that their inclusion is a statement about, first, the general importance of the category of grain crops and, second, the former importance of the archaic grains within this category.

The *padi taun*, along with rice and several other cultigens of great historical and economic importance, are used in the ceremonial feast that marks the end of the harvest and the end of the year's swidden cycle, as well as in a number of other such ceremonies. They constitute the obligatory components of the *pegela'* ("offering") that is made to the spirits during all propitiatory ceremonies.

Discussion and Conclusions

A recurrent theme in these data on crop histories is the notion of historical competition or succession in the political sense of the word.

Contest, History, and Visibility

Many of the mythological, ritual, and cultural mechanisms for representing the succession of crops and cropping systems depict it as a contest.[38] In this representation, the crops are characterized as being "afraid," as "conversing" with one another, as being parent or child, ignorant or greedy, and so on. This is a political vision of crop histories and also, obviously, a very human one. The metaphor of succession links what is with what was; it is a way of constructing indigenous history.

This indigenous construction of history problematizes the relations among crops and crop traditions, but it does not deny them. Vandana Shiva, in a recent (1993) book on biodiversity and biotechnology, draws an analogy between the way people are treated under oppressive political systems and the way indigenous knowledge and biodiversity are treated. Referring to the way that the former military junta in Argentina "disappeared" all real and imagined opponents *(los desparacidos)*, she argues that Western development has tended to treat indigenous cultigens and knowledge the same way (Shiva 1993:9), simply "disappearing" what it deems outmoded. Great losses of indigenous crops and knowledge have occurred in the past half-century because developmental planners either did not take steps to conserve them or because they took active steps to extinguish them. Shiva argues that Western-driven development has pursued what is in effect a "politics of disappearance."

This politics of disappearance is very different from what happens in the indigenous crop histories that have been described here. Far from "disappearing" old cultigens, the Dayak retain them in fields, rituals, language, and myth. Indeed, they have no mechanisms for doing otherwise: Although there are myths, for example, about acquiring new cultigens, there are none about getting rid of old ones. The development paradigm, in contrast, abounds with such myths. An example are the stories spread by European rubber planters during the 1930s concerning the disease and overtapping that purportedly characterized southeast Asian rubber smallholdings, with their unselected rubber varieties and "Jungli" appearance (Dove 1996). These stories, which were used to justify cutting back on and restricting the further expansion of smallholdings, were subsequently shown to be completely without basis. (Indeed, they were shown to be part of an international effort to protect the estate sector from the growth of a more efficient smallholder sector.)

Another example is the belief, prevalent during the early days of the Green Revolution, that the continued cultivation of traditional grains threatened the pest balance with the new high-yielding varieties of grains. In fact, the biggest threat to this balance came from the internal ecological logic of the Green Revolution technology itself (Fox 1993). Nonetheless, the misplaced criticism of traditional grains was invoked to justify such drastic sanctions as burning the fields of the farmers who violated government edicts against their continued cultivation.

The fact that the Dayak can reconcile new and old, past and present, without resorting to the politics of disappearance has important ramifications: Most notably, it reveals the history and thus contingency of contemporary agriculture. For example, the fact that the Kantu' still grow Job's tears and Italian millet in their rice swiddens minimally demonstrates that they can grow these archaic grains, which thus means that they do not have to grow rice (or grow it as exclusively as they now do). This parallel cultivation of new and old cultigens demonstrates the irrelevance of a large number of agronomic variables that might otherwise be invoked to explain why certain agricultural patterns should be brought in or taken out, while driving home the point that agricultural patterns are ultimately explained, in part, by nonagronomic factors. In contrast, the "disappearing" approach of the Western developmental paradigm leaves a false image of an agriculture without history, and thus without alternatives. The denial of alternatives is very much in keeping with the underlying rationale of this developmental paradigm, which is highly deterministic. The tribal pattern retains and celebrates this history, thereby transcending the present.

Transcendence

The stability and sustainability of systems of agriculture like that of the Dayak lie, in part, in the fact that such systems are multidimensional. I refer here to the fact that the Dayak system of agriculture contains multiple "transactional orders" (Bloch and Parry 1989:23–24). One transactional order, which involves the cultivation of rice and other plants of either subsistence or market importance, is focused on the present and on economic concerns.[39] The second transactional order, which involves the cultivation of the archaic grains and other plants of ritual importance, is focused on the past and on ideological concerns. The first involves cultivating rice to meet the subsistence needs of one's own family during the coming year; the second involves cultivating archaic plants to meet the ritual needs of the ancestral spirits of the community.

Although this second transactional order is focused on the past, it is in the present. As a result, as Maurice Bloch says in his 1977 article "The Past and the Present in the Present," the past allows people in the present to talk about the present. This is otherwise difficult because of what Durkheim (1965) called the social determination of social facts, which Bloch (1977:281) summarizes as follows: "If we believe in the social determination of concepts . . . this leaves the actors with no language to talk *about* their society and so change it, since they can only talk *within* it. This is the problem of trying to, in effect, transcend a system while remaining within it." Bateson (1958:302) calls this the problem of "logical typing." He writes: "It is not, in the nature of the case, possible to predict from a description having complexity C what the system would look like if it had complexity C + 1." Bateson (1958:298–99) suggests that the one way to overcome this difficulty is through the use of a language of greater complexity, abstraction, or higher logical type. For Bloch, this language involves a second cognitive system, characterized not by everyday communication but by ritual communication. One salient component of this system is the presence of the past in the present. Dumont (1977:7) also sees this as the solution. He writes (cited in Boon 1982:108): "The main task of comparison is to account for the modern type in terms of the traditional type. For this reason, most of our modern vocabulary is inadequate for comparative purposes: The basic comparative model has to be nonmodern."

The presence of the past in the present—as in the Dayak cultivation of ritual plants—provides, in effect, transcendence: a mechanism by which the social present—with its narrower, shorter term vision—can be transcended. It provides, to return to the theme of this volume, a mechanism by which the classifiers themselves can be *declassified*. The ability to transcend not just the present but also one's own community, culture, or nation is receiving greater attention as worries mount about the state of the global ecosystem. It increasingly appears that the lack of this transcendence is at the root of much of the behavior that imperils our global system. Many resource management fiascoes originate with the sort of parochialism that is the very opposite of this transcendence—something that appears to have been present all along in the agricultural systems of societies like the Dayak.

Relevance

The findings of this analysis offer a new perspective on the ethnography of indigenous history and cultural memory. They suggest, following Bloch's 1977 analysis, that the way that the past is represented in the present can help to address the cognitive challenge of trying to transcend the present while living within it.

The findings of this analysis are also relevant to conservation planning, especially to the current debate about in situ conservation (Altieri and Merrick 1988; Brush 1991, 1992; Oldfield and Alcorn 1989). The present analysis documents an indigenous pattern of what is, in effect, in situ conservation. Moreover, this conservation is being practiced for the right reason: conservation of a historic heritage to place present practice in perspective. (Note that this is not necessarily the same thing as conserving the biogenetic heritage from the past for the future, which is the typical reasoning of contemporary scientists and environmental activists.) This is the type of in situ conservation that development programs should be trying to emulate, in which the conservation is driven by ideology, not the reverse (that is to say, where the perceived need for conservation drives the attempt to develop ideology).

This brings us to perhaps the most important finding of all, regarding sustainability. A question driving many studies of natural resource use today is "What makes some systems of resource use sustainable and others unsustainable?" Research devoted to answering this question has, to date, focused on the material and technological aspects of sustainability. The findings of the present analysis suggest that the ideological aspects of resource use systems may be equally, if not more, important. The transcendent cognitive structure that is often associated with the wise husbandry of resources is likely to be a more important determinant of sustainability than the vast majority of developmental inputs.

Notes

1. Prior to the work of Rosaldo (1980), Wolf (1982), and others in the past decade or so, there was a tendency to view societies on the periphery of the world system as having no histories, but only timeless pasts.

2. Thus, Pelzer (1945:7) and Spencer (1966:114) write as follows: "The folklore and religious practices of many tribes in Borneo and Celebes who today grow cereals recall the time when these tribes had only root crops" (Pelzer). "In many areas where taro and yams were formerly staples, they bear vestiges of ritualism, though the main force of ritualism has been transferred to crops that have more recently become staples" (Spencer).

3. The predominance of rice today is such that to not eat rice is to claim almost nonhuman status (see Tsing [1993:280] on nonconsumption of rice by a Meratus shaman).

4. At the extreme end of this spectrum, in ritual communities found throughout upland southeast Asia whose identity derives from asserting traditional ways of life in the face of contemporary change, archaic plants are in the majority. The Teng-

ger, Badui, and Kasepuhan (Adimihardja 1992) of West Java are examples of such communities.

5. Cf. Dentan's (1971:140) report that the Semai of the Malay Peninsula insist that they have "always" cultivated the new world crops maize and cassava.

6. For the same reason that the historic presence of taro and absence of rice among the early colonizers of the Pacific is thought to attest to the greater antiquity of taro, Harlan (1977:381) suggests that because Job's tears is found in regions still not penetrated by rice, it is thought to predate rice. Even in contemporary centers of rice cultivation like Java, Boomgaard (1989:92) writes: "It was believed around 1800 that either sorghum or millet (or both) had once been more important than rice in Java. However that may be, around 1800 their role must have been rather marginal."

7. Similarly, some of the region's aboriginal tribes may have taken their name from the grain: Skeat and Blagden (1960, 1:341) speculate that the Semang of the Malay Peninsula may have gotten their "nickname," Orang Sekoi (or Sakai), from "the fact of these tribes being millet-eaters," since the Semang term for fox-tail millet is *sekoi*.

8. As Condominas (1972:56) writes, "The adoption of one method of cultivation did not eliminate other forms of cultivation or even other ways of acquiring food-plants."

9. As Fox (1991:253, 256) writes of the island of Roti in Eastern Indonesia: "Although maize, sorghum and other food crops have created a more diverse subsistence basis, they have not altered the ritual focus of Rotinese agriculture on rice and millet.

Despite new and varied dry land cropping patterns in which a variety of European-introduced crops provide the major sources of present-day subsistence, the outer arc of the lesser Sundas retains a heritage embodied in its rituals that still focuses on rice and millet and, to a lesser extent, on Job's tears and sorghum."

10. Nor does it have any importance among the Kantu' as a source of liquor, although neighboring groups use it for this purpose (contra Hill 1977:13). Nor does taro have any ritual importance among the Kantu' and their neighbors (but see Hose and McDougall 1966, 2:8).

11. Note that this myth appears to refer to the early gathering as opposed to the subsequent cultivation and domestication of taro.

12. Similarly, Condominas (1986:40) and Pelzer (1945:8) suggest that the agricultural rituals of the Mnong Gar and the Bontoc reflect, respectively, the general historic precedence of tubers to cereals and the specific precedence of taro to rice.

13. An exception is the delicacy *mpin*, which is made by threshing, winnowing, soaking, husking, and then winnowing again (but not cooking) the new and tender kernels of the not-quite-ripened glutinous rice crop.

14. The practice of roasting tubers (and other foodstuffs) in the ashes of open cook fires is still common among contemporary hunter-gatherers (Stahl 1989:182–83). Pelzer (1945:6–7) suggests that the greater simplicity of preparation of tubers is itself an argument for their historic precedence.

15. In southeast Asia, bamboo containers historically filled part of the niche

subsequently occupied by metal cooking containers (Stahl 1989:183); but observation of the use of bamboo among contemporary Dayak cultivators (and also Penan hunter-gatherers) suggests that it is more suited for the cooking of meat and fish than tubers.

16. This is not the only possible explanation of the charcoal-eating memory, however. There is a clear, symbolic association between this memory and the basic fact of subsistence in a swidden society: Swidden farmers eat crops that are grown on (that is, that "eat") the burnt, carbonized forest in which they make their swiddens. This association is explicit among some swidden groups in the region: Thus, in 1981, highland swidden peoples in Sumbawa (Eastern Indonesia) challenged me with the statement "We eat charcoal!" to see if I understood the dynamics of their land-use system (see Dove 1984). The claim that the ancestors of the Kantu' ate charcoal may, therefore, be no more than an acknowledgment of this fact.

17. This mixture is then thrown at and rubbed on the participants in the ritual, in a variant on the "charcoal-smearing" that was until recently common in traditional ritual contexts in upland southeast Asia.

18. This element of preference is present in many stories in the region about the origins of rice. Thus Psota (1992:32n.2) writes of the Rejang of South Sumatra: "Rice and these two plants [Job's tears and millet] were supplied by the gods to the Rejang, but the Rejang preferred rice."

19. It is also evident in the contemporary Iban proscription against letting any green leaves get into a grave before it is covered. The internment of herbaceous matter would be in effect propagation by vegetative means. If the posited association between tuber mound and human grave is correct, then the corpse takes the role of this herbaceous matter; and the accidental internment of any real herbaceous matter in the grave is undesirable because it would confuse this relationship between signifier and signified. The proscription of such interment thus supports the posited association between graves and root crop mounds. A related meaning is perhaps seen in the contemporary use of taro greens to signify serious illness among the Kayan (Hose and McDougall 1966, 2:8). A possibly related practice in Polynesia is the historic practice (Kirch 1994:277–78) of burying fallen enemy warriors in taro fields.

20. The retort by elders that one occasionally hears is "*tua-tua keladi, semakin tua semakin menjadi*" ("old taro, the older it gets, the better it gets"; cited in *Kedaulatan Rakyat* 1984).

21. Barton (1946:5) argued that the greater uncertainty in rice cultivation, compared with sweet potato cultivation, "explains" why ritual focuses on the former. Brosius (1988), however, suggests that the role of rice production in social reproduction also must be taken into account.

22. See Dove (1983:518).

23. Rice production compares poorly to other sorts of vegetative production as well. For example, Strickland (1986:131) claims that sago production in Borneo is two to four times as energy efficient as rice cultivation.

24. Referring to the nutrient demands of early cultivation, Sauer (1952:89) reminds us of the biblical story of Isaac the cultivator and Esau the hunter. This story

repeats the essential structure of the myth of Pulang Gana, in that it is Esau's failed search for game that leads to his acquiring grain from Isaac.

25. A number of observers have suggested that the great antiquity of Job's tears can be inferred from its distinctive contemporary ritual uses (Condominas 1972:50; Dentan 1991:426).

26. This is reflected in the fact that maize often took its name from Job's tears (Condominas 1972:48, 54). For example, the Kantu' term for Job's tears is *lingkau lesit* "shelling/pecking (nonrice) grain" and their term for maize is *lingkau keribang* "biting (nonrice) grain" (Richards 1981:159, 191, 194). Cf. Fox (1991:250, 254–55) on the way that maize in eastern Indonesia was initially categorized as a type of sorghum, another archaic grain.

27. Hutterer (1983:182n.8) further suggests that Job's tears can be a "volunteer" in swidden fields.

28. This inferred progression from Job's tears and millet to rice is presented here as a general historical pattern from which there have been many local deviations. In northeastern India, for example, Arora (1977:362) reports that Job's tears and rice are grown and in fact cooked (boiled) together, in proportions of one to two. This practice attests both to the similarity in the economic niches that the two grains fill, and to the fact that in this corner of the region the supplanting of the archaic grain has been less complete (or is still in process).

29. For example, Boomgaard (1989:92) writes of Java: "If [the history of] rice is well documented, maize fairly well, and roots and tubers rather badly, virtually nothing is known about a number of minor cereals that must have been of some importance around 1800. I am referring to sorghum or *cantel* (*Sorghum bicolor*), foxtail millet or *juawut* (*Setaria italica*), and Job's tears or *jali* (*Coix lachryma-jobi*). After 1825 these cereals are hardly ever mentioned in the literature or in the archival sources."

30. Cf. Visser (1989:35) on the Toraja: "According to informants, millet and rice were planted together in the beginning of the twentieth century, and even today, one can occasionally see such mixed planting."

31. Schneider (1995:22) recently reported finding Job's tears planted along the boundaries of most rice swiddens in southwest Sumatra. Psota (1992:32n.2) similarly writes that Job's tears (plus some sort of millet) "are often planted at the edge of the hill rice fields" in South Sumatra. Barrau (1965b:292n.2) earlier reported the cultivation of Job's tears in New Guinea for food, a source of ash-salt, and ornament. Watt (1966:397), however, notes that cultivation of Job's tears is detrimental to the ornamental value of the grain.

32. Cf. Dentan (1991:426), who reports that the Semai, who cultivate and use Job's tears today only for ornamental purposes, formerly used it in their thunder squall ritual and remember it from before that as the ancestor of their crops.

33. A variant belief holds that the archaic grains do not threaten but protect the rice (viz., from conflict with others; Cramb 1985:40).

34. Similarly, Schneider (1995:111) relates a myth from southwest Sumatra, in which the gods first give humans millet, then Job's tears, and finally rice. Condominas (1986:40) also cites ritual evidence for the historic precedence of Job's tears to rice.

35. The fact that Job's tears, millet, and rice can not only all be planted within the same general type of field but also in the same type of soil within that field [Spencer 1966:35], would only exacerbate this perceived tension among them.

36. The postulated history of crop succession has left behind a proscription about the planting distance between Job's tears and foxtail millet, but not about the planting distance between either one and rice, because rice is now too dominant and ubiquitous a cultigen for any distance at all to be possible.

37. But see Ochse's (1980:56) report of a Javanese belief that tubers cause root-rot in rice.

38. This is a common idiom: For example, the early-twentieth-century confrontation between swidden rice production and rubber tapping in Borneo was expressed among the Dayak of Borneo in a dream in which the rubber trees "eat" the spirit of the rice (Dove 1996).

39. Within this transactional order, a further distinction can be seen between the market and individually oriented order on the one hand, and a subsistence and group-oriented order on the other (Dove 1996).

References

Adimihardja, Kusnaka. 1992. "The Traditional Agricultural Practices of the Kasepuhan Community of West Java." Pp. 33–46 in *The Heritage of Traditional Agriculture among the Western Austronesians*. James J. Fox, ed. Occasional paper, Department of Anthropology, Australian National University, Canberra.

Altieri, Miguel, and Laura C. Merrick. 1988. "Agroecology and In Situ Conservation of Native Crop Diversity in the Third World. Pp. 361–69 in *Biodiversity*, E. O. Wilson, ed. Washington, D.C.: National Academy Press,

Arora, R. K. 1977. "Job's-tears (*Coix lacryma-jobi*)—A Minor Food and Fodder Crop of Northeastern India." *Economic Botany* 31:358–66.

Barrau, Jacques. 1965a. "L'humide et le sec: An Essay on Ethnobiological Adaptation to Contrastive Environments in the Indo-Pacific Area." *Journal of the Polynesian Society* 74:329–46.

———. 1965b. "Witnesses of the Past: Notes on Some Food Plants of Oceania." *Ethnology* 4(3):282–94.

Barton, Roy F. 1946. "The Religion of the Ifugao." Memoir no. 65. *American Anthropologist* 48:1–219.

Bateson, Gregory. 1958. *Naven*. Stanford: Stanford University Press.

Bellwood, Peter. 1979. *Man's Conquest of the Pacific: The Prehistory of Southeast Asia and Oceania*. New York: Oxford University Press.

Bloch, Maurice. 1977. "The Past and the Present in the Present." *Man* 12:278–92.

Bloch, Maurice, and Jonathan Parry. 1989. "Introduction: Money and the Morality of Exchange." Pp. 1–32 in *Money and the Morality of Exchange*. J. Parry and M. Bloch, eds. Cambridge: Cambridge University Press.

Boomgaard, Peter. 1989. *Children of the Colonial State: Population Growth and Economic Development in Java, 1795–1880*. Amsterdam: Free University Press.

Boon, James A. 1982. *Other Tribes, Other Scribes: Symbolic Anthropology in the Comparative Study of Cultures, Histories, Religions, and Texts*. Cambridge: Cambridge University Press.

Brosius, J. Peter. 1988. "Significance and Social Being in Ifugao Agricultural Production." *Ethnology* 27(1):97–110.

Brush, Stephen B. 1991. "A Farmer-based Approach to Conserving Crop Germplasm." *Economic Botany* 45:153–66.

———. 1992. "Farmers' Rights and Genetic Conservation in Traditional Farming Systems." *World Development* 20(11):1617–30.

Burkill, I. H. 1951. "The Rise and Decline of the Greater Yam in the Service of Man." *Advancement of Science* 7:443–48.

———. [1935]. 1966. *A Dictionary of the Economic Products of the Malay Peninsula*. 2 vols. Kuala Lumpur: Ministry of Agriculture and Cooperation.

Chesnov, Ia. V. 1973. "Domestication of Rice and the Origin of Peoples Inhabiting East and Southeast Asia." Paper presented at the ninth International Congress of Anthropological and Ethnological Sciences, Chicago, August–September 1973.

Condominas, Georges. 1972. "From the Rice-field to the Mir." *Social Science Information* 11(2):41–62.

———. 1986. "Ritual Technology in Mnong Gar Swidden Agriculture." Pp. 28–46 in *Rice Societies: Asian Problems and Prospects*. Irene Nørlund, Sven Cederroth, and Ingela Gerdin, eds. Curzon, London, and Riverdale, Md.: Scandinavian Institute of Asian Studies.

Conklin, Harold C. 1959. "Ethnobotanical Problems in the Comparative Study of Folk Taxonomy." Proceedings of the ninth Pacific Science Congress, vol. 7: 299–301.

Coursey, D. G., and C. K. Coursey. 1971. "The New Yam Festivals of West Africa." *Anthropos* 66:444–84.

Cramb, R. A. 1985. "The Importance of Secondary Crops in Iban Hill Rice Farming." *Sarawak Museum Journal* 34(55): 37–45.

Crevost, Charles, and Charles Lemarie. [1901–2]. 1917–1941. *Catalogue des produits de l'Indochine*. Hanoi: Extreme-Orient.

Dentan, Robert Knox. 1968. *The Semai: A Nonviolent People of Malaya*. New York: Holt, Rinehart and Winston.

———. 1971. "Some Senoi Semai Planting Techniques." *Economic Botany* 25:136–59.

———. 1991. "Potential Food Sources for Foragers in Malaysian Rainforest: Sago, Yams and Lots of Little Things." *Bijdragen* 147(4):420–44.

Dove, Michael R. 1983. "Review of H. C. Conklin's Ethnographic Atlas of Ifugao and Its Implications for Theories of Agricultural Evolution in Southeast Asia." *Current Anthropology* 24(4):516–19.

———. 1984. "Man, Land, and Game in Sumbawa: Some Observations on Agrarian Ecology and Development Policy in Eastern Indonesia." *Journal of Tropical Geography* 5(2):112–24.

———. 1985a. "The Agroecological Mythology of the Javanese, and the Political Economy of Indonesia." *Indonesia* 39:1–36.

———. 1985b. *Swidden Agriculture in Indonesia: The Subsistence Strategies of the Kalimantan Kantu'*. Berlin: Mouton.

———. 1996. "Rice-eating Rubber and People-eating Governments: Peasant versus State Critiques of Rubber Development in Colonial Indonesia." *Ethnohistory* 43(1):33–63.

Dumont, Louis. 1977. *From Mandeville to Marx*. Chicago: University of Chicago Press.

Durkheim, Emile. [1912]. 1965. *The Elementary Forms of the Religious Life*. Joseph Ward Swain, trans. New York: Free Press. [English edition of *Les formes elementaires des la vie religieuse: Le systeme totemique en Australie*, published by Alcan, Paris.]

Fox, James J. 1991. "The Heritage of Traditional Agriculture in Eastern Indonesia: Lexical Evidence and the Indication of Rituals from the Outer Arc of the Lesser Sundas." *Indo-Pacific Prehistory Association Bulletin* 10:248–62.

———. 1993. "Ecological Policies for Sustaining High Production in Rice: Observations on Rice Intensification in Indonesia." Pp. 211–24 in *Southeast Asia's Environmental Future: The Search for Sustainability*. Harold Brookfield and Yvonne Byron, eds. Tokyo: UNU Press; Kuala Lumpur: Oxford University Press.

Harlan, Jack R. 1977. "The Origin of Cereal Agriculture in the Old World." Pp. 357–83 in *Origins of Agriculture*. Charles A. Reed, ed. The Hague: Mouton Publishers.

———. 1992. *Crops and Man*. 2d ed. Madison, Wisc.: American Society of Agronomy and Crop Science Society of America.

Harris, David. 1972. "The Origins of Agriculture in the Tropics." *American Scientist* 60:180–93.

———. 1973. "The Prehistory of Tropical Agriculture: An Ethnoecological Model." Pp. 391–417 in *The Explanation of Culture Change: Models in Prehistory*. C. Renfrew, ed. Pittsburgh: University of Pittsburgh Press.

———. 1977. "Alternative Pathways toward Agriculture." Pp. 179–243 in *Origins of Agriculture*. Charles A. Reed, ed. The Hague: Mouton Publishers.

Hill, R. D. 1977. *Rice in Malaya: A Study in Historical Geography*. Kuala Lumpur: Oxford University Press.

Hose, Charles, and William McDougall. [1912]. 1966. *The Pagan Tribes of Borneo*. 2 vols. Reprint: London: Frank Cass & Co.

Hutterer, Karl Leopold. 1983. "The Natural and Cultural History of Southeast Asian Agriculture." *Anthropos* 78:169–212.

———. 1988. "The Prehistory of the Asian Rain Forests." Pp. 63–72 in *People of the Tropical Rain Forest*. Julie Sloan Denslow and Christine Padoch, eds. Berkeley: University of California Press; Washington, D.C.: Smithsonian Institution.

Kedaulatan Rakyat (daily newspaper, Yogyakarta). 1984. "Tua Keladi" ("Old Tuber/ Taro"). July 19, page 7.

Kirch, Patrick Vinton. 1994. *The Wet and the Dry: Irrigation and Agricultural Intensification in Polynesia*. Chicago: University of Chicago Press.

Li, Hui-lin. 1970. "The Origin of Cultivated Plants in Southeast Asia." *Economic Botany* 24(1):3–19.

Ochse, J. J., in collaboration with R. C. Bakhuizen Van Den Brink. [1931]. 1980. *Vegetables of the Dutch East Indies*. Amsterdam: A. Asher and Co. [English edition of *Indische Groenten*, published by Archipel Drukkerij, Buitenzorg.]

Oldfield, Margery L., and Janis B. Alcorn. 1989. "Conservation of Traditional Agroecosystems: Can Age-Old Farming Practices Effectively Conserve Crop Genetic Resources?" *Bioscience* 37(3):199–208.

Pelzer, Karl J. 1945. *Pioneer Settlement in the Asiatic Tropics: Studies in Land Utilization and Agricultural Colonization in Southeastern Asia*. New York: American Geographical Society.

Psota, Thomas M. 1992. "Forest Souls and Rice Deities": Rituals in Hill Rice Cultivation and Forest Product Collection." Pp. 30–51 in *The Rejang of South Sumatra*. University of Hull Occasional Paper no. 19.

Raffles, Thomas Stamford. [1817]. 1978. *The History of Java*. 2 vols. Kuala Lumpur: Oxford University Press. [Originally published by Black, Parbury, and Allen, London.]

Richards, A. 1981. *An Iban-English Dictionary*. Oxford: Clarendon Press.

Rival, Laura. 1993. "The Growth of Family Trees: Understanding Huaorani Perceptions of the Forest." *Man* 28(4):635–52.

Rosaldo, Renato. 1980. *Ilongot Headhunting 1883–1974: A Study in Society and History*. Stanford: Stanford University Press.

Roth, Henry Ling. [1896]. 1980. *The Natives of Sarawak and British North Borneo*. 2 vols. London: Truslove and Hanson. [Reprinted by University of Malaya Press, Kuala Lumpur.]

Sather, Clifford. 1977. "*Nanchang Padi*: Symbolism of Saribas Iban First Rites of Harvest." *Journal of the Malaysian Branch of the Royal Asiatic Society* 50(2):150–70.

Sauer, C. O. 1952. *Agricultural Origins and Dispersals*. New York: American Geographical Society.

Scaglion, Richard, and Kimberly A. Soto. 1994. "A Prehistoric Introduction of the Sweet Potato in New Guinea." Pp. 257–94 in *Migration and Transformations: Regional Perspectives on New Guinea*. A. Strathern and G. Sturzenhofecker, eds. Pittsburgh: ASA Monographs no. 4.

Schneider, Jürg. 1995. *From Upland to Irrigated Rice: The Development of Wet-rice Agriculture in Rejang Musi Southwest Sumatra*. Berlin: Reimer.

Scott, N. C. 1956. *A Dictionary of Sea Dayak*. London: School of Oriental and African Studies, University of London.

Shiva, Vandana. 1993. *Monocultures of the Mind: Perspectives on Biodiversity and Biotechnology*. Dehra Dun, India: Natraj Publishers.

Skeat, W. W., and C. O. Blagden. [1906]. 1960. *Pagan Races of the Malay Peninsula*.

2 vols. London: Macmillan and Co. [Originally published in London by Macmillan and Co.]

Spencer, J. E. 1966. *Shifting Cultivation in Southeastern Asia*. Berkeley: University of California Press.

Stahl, Ann B. 1989. "Plant-food Processing: Implications for Dietary Quality." Pp. 171–94 in *Foraging and Farming: The Evolution of Plant Exploitation*. David R. Harris and Gordon C. Hillman, eds. London: Unwin Hyman.

Strickland, S. S. 1986. "Long Term Development of Kejaman Subsistence: An Ecological Study." *Sarawak Museum Journal* 36(51):117–71.

Sugishima, Takashi. 1994. "Double Descent, Alliance, and Botanical Metaphors among the Lionese of Central Flores." *Bijdragen* 150(1):146–70.

Tsing, Anna L. 1993. *In the Realm of the Diamond Queen: Marginality in an Out-of-the-Way Place*. Princeton: Princeton University Press.

Visser, Leontine E. 1989. *My Rice Field Is My Child: Social and Territorial Aspects of Swidden Cultivation in Sahu, Eastern Indonesia*. Rita DeCoursey, trans. Verhandelingen #136. Dordrecht: Foris Publications.

Watt, Sir George. [1908]. 1966. *The Commercial Products of India: Being an Abridgment of "The Dictionary of the Economic Products of India."* Faridabad: Today and Tomorrow's Printers and Publishers. [Originally published by His Majesty's Secretary of State, New Delhi.]

Wolf, Eric R. 1982. *Europe and the People without History*. Berkeley: University of California Press.

Yen, Douglas E. 1971. The "Development of Agriculture in Oceania." Pp. 1–12 in *Studies in Oceanic Culture History*, vol. 2. R. C. Green and M. Kelly, eds. Pacific Anthropological Records 12. Honolulu: Bishop Museum Press.

———. 1975. "Indigenous Food Processing in Oceania." Pp. 147–68 in *Gastronomy: The Anthropology of Food Habits*. Margaret L. Arnott, ed. The Hague: Mouton.

5

Ethnoecology Serving the Community

A Case Study from Zuni Pueblo, New Mexico

RICHARD I. FORD

Ethnoecology has an interesting history and a critical role as anthropology evolves into a policy-relevant social science. It draws upon a number of disciplines and integrates them into a comprehensive methodology that has general application for solving problems and seeking an understanding of community adaptation to natural environmental factors and to a matrix of asymmetrical social forces. The emerging ethnoecological methodology is a powerful tool for engaged anthropology.

Ethnoscience includes ethnoecology and a series of additional "ethno-" subjects that generally are defined as scientific according to Western definitions. These include ethnobotany, ethnogeography, and ethnozoology, among others. Ethnoscience grew out of detailed linguistic descriptions of cultures (Sturtevant 1964). Many of these studies have contributed to a growing awareness in anthropology and psychology that general principles of cognition can be derived through comparative studies of the results of linguistic research. Berlin's *Ethnobiological Classification* (1992) is the most salient volume to generalize from these studies and to provide a foundation for further research.

Several early papers by Conklin (1957) and Frake (1962) pointed anthropological ecology in the direction of the incorporation of ethnoscientific details by studying specific cultures. Unfortunately, from a historical perspective, their pioneering efforts were virtually disregarded as anthropologists simultaneously discovered the work of scientific ecologists and applied their terminology, units of analysis, and methodology to understanding the adaptations of human populations. As a consequence, ecological investigations improved over the earlier studies, but they were skewed by the absence of native perspectives and their rules about adapting to the environment.

Within American anthropology, at the same time that linguistic study was demonstrated to have ecological consequences, cultural ecology was proceeding in new directions, separate from that advocated by Julian Steward and Leslie White (Orlove 1980). The ecological anthropology advanced by Andrew Vayda and Roy Rappaport (1968) and by Rappaport (1968) encouraged analyses drawn from the general ecology of living organisms in ecosystems. But for them, ethnoscience had a place as part of the "cognized environment" (Rappaport 1965) or "the sum of the phenomena ordered into meaningful categories by the population" (1965:159). Rappaport contrasted this term with the "operational environment" (1965:159)—that is, the phenomena that impinge on the well-being of a population as determined by outside analysts. Some elements of the two environments may overlap, while others may go unrecognized by both the natives and the researcher.

The importance of the redefinition of anthropological ecology led to some interesting studies that had few antecedents in the understanding of the ecological behavior of ethnographic populations. The result was several studies that depended upon knowledge of native classifications to understand their relations with nature (see, for example, Johnson 1974).

The goals and initial results of this neofunctionalist ecology were critiqued and relegated by some to another form of ethnographic description. Even some of the founders of the "new ecology" desired to set it on a new course. Vayda (Vayda and McCay 1975) focused attention on environmental problems and responses to them by individuals (rather than populations) as the unit of concern. At that point, ecological anthropology became a field for analyzing ecological disasters, their multiple social impacts, and the solutions that could potentially resolve them. This position spoke for anthropology as a voice in public policy issues, coinciding with an increase of environmental awareness that found a new forum in American culture.

This should have been the appropriate time for ethnoscience to assert itself as a critical component of ecological studies, but it did not. References to ethnoscience are not found in an important overview of ecological anthropology published in 1980 (Orlove). It is only in a chapter by Fowler (1977) in a leading ecological anthropology textbook written in the same decade (Hardesty 1977) that ethnoecological concepts and approaches are discussed.

This brief summary reveals that ethnoecology was well established (Conklin 1957 and Fowler 1977) and that ecological anthropology had a place in the solution of environmental problems, but that there was not an established methodology to merge them. More recent studies and some experimental endeavors have evolved an ethnoscience methodology that underlies ethnoecological investigations.

Although the published studies are not numerous, those that are available are extremely instructive. J. Stephen Lansing (1991) has looked beyond individual populations at the farmers who derive water from centralized distribution centers (water temples) in Bali. He had to learn their classification of spatial terms, social hierarchies, and beliefs about water before he could begin to understand the importance of ritual for solving a fundamental problem of survival. However, ethnoecology alone would not answer his questions, and he had to analyze the system from a hydrologic perspective and combine the different viewpoints into a computer simulation.

Furthermore, much of the ethnoscience literature had gender and social biases favoring male native intellectuals well versed to provide answers about classifications of natural phenomena. Virginia Nazarea corrected these deficiencies in her studies by exploring the range of variation in knowledge about sweet potatoes (1991) and rice (1990, 1995) by gender, age, and social class in the Philippines. Her work has been an important contribution to the development of an ethnoscience methodology appropriate for ethnoecological concerns.

Ethnoscience as a methodology must include the cognitive environment and the operational environment, to resurrect Rappaport's terms. However, what questions will be asked and what variables will be recorded depend upon the problem under investigation. Most ecological studies in anthropology are no longer based on an eighteen-month residence in a community by a single anthropologist who describes local adaptation(s). They now may be shorter in duration and may often involve a team of investigators, and the results are aimed at finding a solution to a problem for a community or group who are of limited power in the matrix of political relations in which they are situated.

Zuni Pueblo: A Case Study

I was asked to serve on a legal team as an expert witness in behalf of Zuni Pueblo and its litigation against the U.S. government for desecration of trust land. The attorney for the pueblo, Stephen Boyden, from Salt Lake City, assembled a task force consisting of an archaeologist, a geologist, a historian, a photometrician, a ranchland adjuster, and an ethnobotanist. I was the ethnobotanist. We first had to meet with the elected Zuni tribal council and then with Zuni citizens at a town meeting. Upon endorsement by both groups we could conduct research directly related to the case, interview willing individuals, and photograph the lands once used by the Zuni without having to secure further permission. Henceforth, I will use the term "Zuni" to refer both to the land and to the people.

At first glance, Zuni presents a spectacular landscape. It satisfies all the stereotypes one has of the Wild West, including the tumbleweed, and offers the aesthetic pleasure most Anglos derive from a rugged landscape. But behind the tourists' vistas is a badly desiccated land with scars from resource extraction and from the destruction of archaeological sites caused by road construction. The Zuni claimed that these environmental atrocities were the fault of the U.S. government over a century and more of expropriation and misguided land policies. The Zunis asserted, and official archival documents confirm, that the U.S. government mined coal for the railroad that passed through Gallup, New Mexico, and that timber was harvested without proper compensation or concern for the land. The government also deliberately built dams with the intention of changing methods of farming and residence patterns, without consulting the Zuni. The dams in particular caused the erosion that ruined much traditional farmland.

In response, the government admitted no guilt. It argued that its policies were just and were intended to improve the lives of the Zuni. Furthermore, it maintained that erosion is one of the natural forces in the West and "an act of God," not a result of governmental actions.

As the ethnobotanist on the team, I was responsible for determining the traditional farming practices of the Zuni, the plants that they used, and how they herded their livestock. Early into the research, it was apparent that the Zuni language was still spoken by many elderly and middle-aged men and women, and that most were also fluent in English. Members of these age cohorts could help me to learn the appropriate terminology and classifications needed for the study. My initial tour of the reservation and a few preliminary interviews revealed that the portrayal of Zuni Pueblo by anthropologists masks the fact that there are actually five farming villages—Nutria (Upper and Lower), Pescado (Upper and Lower), Tekapo, Ojo Caliente, and Halona:wa. In each, the many farming practices are different.

Moreover, extensive field investigations were mandatory, but because of appropriate gender relations I could not conduct them in the field with women. I had to hire a mature female assistant who would interview the women. It was also apparent from published ethnographies and our initial interviews that the clan system remains very important to the Zuni. Only certain adult members of the clans, however, knew the details concerning shrines and other areas that had been affected by the degradation of the reservation. To complicate the social organization further, Zuni has many active sodalities with special rituals and locations for collecting the plants used in ceremonies; only a few initiated men know about these. Thus, to be thorough, this study had to consider the knowledge of farmers in all Zuni villages, the different ecological details known to men and women, and the special lore that membership in certain clans or sodalities bestows upon relatives and members. Over two years the members of the legal team interviewed sixty-four Zuni tribal members about land use and ecological relations as perceived by the Zuni.

Any study of land use practices at Zuni benefits from the work of Frank Hamilton Cushing and his classic *Zuni Breadstuff* (1920). He anticipated ethnoscience by providing the Zuni names of natural phenomena and information on the agricultural practices at the farming village of Nutria. Cushing reported what we demonstrated in our reservationwide study: The Zuni have always been concerned not with soil erosion but with water management, to ensure adequate amounts for their crops. His work gave us invaluable leads on a culture that has no words for erosion or conservation. How did they preserve the land, when the government could not?

In this ecological study, Zuni beliefs about the land and water, their names and taxonomies, and their rules of behavior toward natural resources had to be determined. The environmental consequences of their actions had to be compared with the results of Western ecological methods. Plants were collected and named by Zuni consultants and then identified according to Western systematics. Water flow was determined along Zuni drainage systems and measured. Eroded areas once farmed by the Zuni, as confirmed by the archaeologist, archival photographs, and boundary survey documents, were examined for soil loss, and the geological cause of that loss was assessed. The Zuni had an important role in presenting their science, and their ethnographical statements were corroborated for the court with archaeological survey records, nineteenth-century government and missionary reports, and turn-of-the-century photographs of the area.

Land Use

According to the Zuni, Earth Mother has many rocks, minerals, and soil on her back. For a study of land use, Zuni soil topology had to be understood. Based on our research and observations, we established the following Zuni soil types:

1. Sand *(so:we)*. This soil is regarded as outstanding for corn, beans, and squashes. In drainage bottoms with high subsurface water tables, it supports cornfields. Where sand is deep, squash or beans may be given preference over corn. Peach orchards are planted on dunes.
2. Loam *(helvalo:we)*. This soil is ideal for corn and is usually found in tributary valleys, the bottoms of major drainages, and the outwash fans of seasonal streams.
3. Clay *(hepecha)*. Clay is found in patches in the farming villages, irrigated land, and in greatest abundance in the Zuni Irrigation Project. The cohesiveness of the minerals makes the soil "tight" and more difficult to cultivate using traditional methods of hoe and digging stick agriculture, and it makes it difficult for germinating seeds to penetrate the surface. This soil type supports a luxuriant growth of weeds, requiring extensive work to remove. Wheat is generally grown on this soil and, more recently, alfalfa; it is the least preferred soil for corn. The Zuni believe that corn is more difficult to grow on this soil because of the labor required, and the potential for greater water runoff. The yield from this soil type is also lower than those from the other types.
4. Adobe. This soil is a thick clay and is not used for agricultural fields. Occasionally, patches of this soil are mixed with animal manure for use in gardens *(hekkowetsanna)*.

Water Management

From the Zuni perspective, soils and topography determine the appropriate water control practice. The Zuni employ many agricultural methods, so that they do not have to waste water, but most of these had not been described previously in the literature.

Rainfall Farming

In rainfall agriculture (*kwakowtome* ("water from the sky") or *e'amakwinishe' make deachinen* ("depending on rain field"), three types of farming practices can be recognized as traditional methods of bringing rainwater to planted seeds and growing crops. The purpose of each is to conserve as much water as possible for the plants. Over time, a single field may be treated with one or several methods to maximize available moisture. Observers throughout the American contact period have noted the predominance of rainfall agriculture in Zuni. Cushing stressed the control of rainfall in the outlying farm fields.

Each field cultivated by the men had a name based on the plant it supported, regardless of the system of water control. *Deachinawa* means any kind of planted fields, *duteachinwa* is a cornfield, *chuteachunane* is a field with ripe corn, and *noteanchinwa* is a bean field.

Dry Farming (soyalthlo deachi *["sand on top field"]*)

In this form of agriculture, water from above falls on a surface that is fairly flat and is cultivated so as to retard runoff. Consequently, the rainwater or snowmelt will remain in the field, requiring no further maintenance techniques.

Runoff Farming

Methods have been employed to capture rainwater running down a slope. The Zuni recognize that the great concentration and velocity of water caused by a slope would result in loss of water and possibly damage to crops and the land. Therefore, various methods have been employed to slow the water's progress and to spread it out evenly over the fields or, if necessary, beyond the field borders.

Cushing illustrates one method for controlling runoff. It consists of several check dams of earth and then additional earth baffles to separate the flow and to distribute it evenly to the crop. Traditional farmers describe similar methods today. For example, a V of branches covered by a soil embankment is used to head off the flow of water and thereby reduce its volume and direct its course.

Permanent structures were sometimes made, with stakes set in the soil with brush and stone faces or with stone alignments to interrupt the flow down the slope. In other places, earth dikes would be constructed annually as a more efficient means of controlling runoff. In Nutria, logs of cottonwood or ponderosa pine were left at the edges of fields and then dragged by hand or with horses to intercept runoff and to spread the water along the sides of the heavy logs to the fields.

Floodwater Farming

This is a general term encompassing several techniques for flooding fields with large volumes of water. These fields benefited from both the water and new soil.

The Zuni method of flood-control farming was described in general terms by Stewart (1940), who was senior soil conservationist in the U.S. Soil Conservation Service. He observed that Zuni fields were surrounded by a dirt ridge, eighteen to twenty-four inches high, formed by hand or, more recently, with a plow. This *hekkowethlanna* would catch and retain all water coming into the field. To transfer the water from streams to the corn patches, he noted that the "Zuni cultivator finds it easier to deflect the flood water out of the stream bed with a series of small brush-and-earth dams. . . . A series of herringbone, radiating earth checks, extending out from the stream bank across the stream itself, deflect the water from the channel and spread it fairly evenly over the entire field with the help of the earth checks" (Stewart 1940:335).

Arroyo openings are locations for fields of traditional crops. Arroyos carrying water are partially dammed with stones to slow the flow and to keep it in its channel. When an arroyo opens onto land of a lesser slope, the water is spread across the field by means of a ditch (dug by hand or cut with a plow), creating a flow perpendicular to the arroyo. Water is then released from the ditch across the field in a flood fashion.

The second technique involves locating fields adjacent to arroyos and then damming the ephemeral stream with posts and brush, in effect damming the channel and raising the water onto the fields. The bottom of the arroyo fills with soil, causing the arroyo to disappear over time.

The third method consists of flooding large fields with discharge from the Zuni River and other major drainages. Here check dams are installed to force the water over the bank and make it flow across the land, with water being directed by earth embankments. The major potential problem is damage to crops caused when insufficient manpower prevents control of all the water.

The process of water control is dynamic. The Zuni do not name each type, other than in descriptive terms, because the ultimate objective is to bring water to growing crop plants. For example, a field may be planted at the mouth of an arroyo. A floodwater technique may be used to lift the water over the bank to the field, and in the process a dam made of posts, stone, and brush will cause sediment to accumulate behind it. As the arroyo fills in, the slope across a field may decrease and runoff farming embankments may be all that are needed to spread the water. Finally, within a few years a level field is created and only rain "from above" will be directed to plants. This progression of water control methods is recognized throughout the Zuni reservation.

Outlying farms away from the irrigated lands in Zuni and the farming villages were extensive. A map based upon the personal recollections of a sample of Zunis who farmed before 1946 reveals the extent of this practice and the importance of dry farming for Zuni subsistence and economy.

*Irrigation Farming (*k'va:ti'teyachin'a *["irrigated land"])*

The transport of water in canals from springs and reservoirs characterized the farming villages. The Spanish observed aboriginal canals in Zuni villages in the sixteenth century. However, they were not the most common agricultural method, and only a fraction of the Zuni farmland was so irrigated. Before the Spanish arrived, some corn, beans, and squash were irrigated; after the Europeans came, the major irrigated crop was wheat. In this century, the most significant irrigated crops have been wheat and alfalfa, along with some oats, rye, and barley.

Spring-fed irrigation is best illustrated at Ojo Caliente, where a clear, deep spring, Doseluna, is thirty-five to forty feet long and eighteen to twenty feet wide. Two ditches led from it, with the larger of the two watering a 2.5-by-3.5-mile area. The smaller ditch was higher and the other had to be dammed to lift water into it (Stevenson 1904:351).

The essence of Zuni ceremonialism is the bringing of water—rains and snow. The objective of the water control systems is the conservation of runoff moisture—from the sky, from streams, from springs. In the farming villages, the traditional irrigation system conserved water behind small dams. No special ceremonies were necessary to open the ditches in the spring or close them in the fall: The presence of water was already an answer to their prayers.

Gardens and Orchards

In Zuni, field labor has traditionally been male labor, while women have worked the gardens. The gardens are used to grow a variety of crops, some of which are Spanish-introduced or recent American vegetables. Others are traditional plants and even encouraged "wild" plants. At the turn of the century, women planted coriander, onions, and garlic in their gardens. Some also grew small amounts of cotton for husbands and brothers involved in ceremonial activities. In addition, they encouraged certain wild plants, such as amaranth for red dye (it may have been planted at one time from seed), husk tomatoes, and edible greens. Larger gardens sometimes contained a few bean and corn plants. Within the twentieth century, the crop diversity has expanded with vegetable seeds distributed by the U.S. government or purchased from trading posts and stores.

Pot Irrigation

The classic "waffle" gardens, or *thlateƙwine*, in Zuni Pueblo were walled with adobe or fenced with upright posts, which were sometimes mud covered, to keep livestock, dogs, and children away from the plants. The gardens were small squares, twelve to eighteen inches on a side, arranged in paired rows with pathways between so a woman could water each square from the side. Water was obtained in a ceramic pot or metal container from walk-in wells in the village, from the Zuni River stream flow itself, or, when the river was dry, from holes dug in the sand that filled with water from beneath *(heƙƙonne)*. In July of 1890, Fewkes observed that the bed of the Zuni River was dry except for a small trickle. "Almost at any time one can find water by digging a few feet below the surface, and water obtained this way is ordinarily used in watering the gardens to the west of the pueblo" (Fewkes 1891:37). A gourd dipper or cup was used to ladle water onto the growing plants.

Ditch Irrigation

Gardens are also maintained at the farming villages and on individual ranches. Because they lack upright wells, they are called *heƙƙowe*; it is misleading to refer to them as waffle gardens. These pen gardens are built directly next to the irrigation ditches that extend from springs or dams to the farm fields. The water is channeled through small feeder canals, which the women clean and maintain, from the larger ditch into the garden. These outlying gardens are larger than those in the pueblo and often are rebuilt each year by plowing up the previous year's area and reforming the squares by hand. They are five or ten times larger than the waffle gardens; however, the same plants are raised in them.

During the growing season, wild plants also grow. All recognize the dispersal of seed from plants growing there the previous year. But when a new plant appeared, or to account for the initial appearance of a plant, the source was attributed to seeds dropped with the rain from the sky by the *'uwanani* ("rainmaker"). These seeds grow into *hawe* ("weeds")—that is, plants that are useless, or into wild but useful plants *(ƙ'ute:pathlto)*. A weed *(hale)* is hoed from a garden, but the *ƙ'ute:pathlto* are permitted to grow and are collected for food or other uses. Wild husk tomatoes (*Physalis* sp.), beeweed or *'atowe (Cleome serrulata)*, wild potatoes *(Solanum jamesii)*, amaranth, and many others are cultivated and harvested along with domesticated plants.

Spring Irrigation

Water from natural flowing springs is impounded and led through earthen ditches or wooden logs to small gardens. Men maintain the ponds and ditches, but mothers or wives farm the garden plots, planting the seeds and harvesting the produce. Coriander, onions, and garlic form the principal crops, but cabbage, chili, lettuce, and radishes were also planted.

*Orchards (*mo:chekwatana:yewe*)*

Peach orchards, which are family-owned and inherited through mothers to daughters, were planted at several locations at Zuni. The largest were in the northeast sand dunes on Dowa Yalanne, where subsurface seepage could effectively water the trees. Here small houses were constructed and occupied primarily at the time the peaches ripened.

Range Land

In an effort to enhance grassland for livestock, the Zuni sometimes "brush up"—that is, place brush in—arroyos that have started to damage rangeland. These limbs of juniper, oak, and piñon hold the soil and retard the flow of water by permitting it to filter slowly through the tree branches before heading downstream. Dams are built across the Zuni River west of the pueblo to flood ditches and bring water to grassland. Damming natural stream channels provides water for livestock.

Burning to increase grass production is an old Zuni technique. Fire has long been a means of clearing a field of dry weeds and brush, and the same method has been transferred to grass production for range animals. Early spring is the time preferred by the Zuni as well as advocated by the Bureau of Indian Affairs. James Enote, who is the current Zuni conservationist, notes:

> "[Controlled burning] is a big tool, because it is cheap. It doesn't cost much to go out and get a fire going. When it is done properly, you can get a good kill on the . . . shrubs, target species, . . . especially sagebrush and snakeweed, whereas you can't do too much about the rabbitbrush unless the fire gets really hot. March is a particularly good time because the grasses are, typically, usually still in dormancy. . . . You can't do too much damage with a quick burning fire which won't hurt the growing parts of the grass. Also, the

> above-ground parts of the grass are dried. They provide a . . . fine fuel between the target plant species. And in March, you usually start getting the winds. So that's an ideal time.

Rangeland could be modified into farmland to provide a source for collected plants. The Zuni made extensive use of plant products, with each area of the reservation providing some plant for food, medicine and ceremonies, construction, and fuel. In fact, collecting—*e'tonashikana:ne*, ("go and pick food")—was so important to daily subsistence that until very recently every portion of the reservation supported some plant of use to the Zuni. The intensive collection of edible seeds goes beyond economic necessity. The wild plant seeds are mentioned in prayers and are carried in bags around the waists of masked dancers and in ceremonial wands used in the Shalako ceremony. One published prayer enumerating these wild plants appears in the Sayatasha Kachina's night chant at Shalako (Bunzel 1932:714–15).

Government Intervention

The Black Rock Dam

Between 1906 and 1909, one of the largest public works projects in the United States, Black Rock Dam, was completed at Zuni Pueblo. It blocked the Zuni River in a canyon of volcanic basalt and created a large lake. Black Rock Dam was built to provide irrigation for a large section of land north of the present pueblo. The land was sectioned into individual allotments that were fed by feeder canals from the main canal.

The genesis of the idea to build a dam on the Zuni reservation to impound water for irrigation cannot be attributed to the Zuni people. E. Richard Hart has thoroughly scrutinized the federal archives for evidence of Zuni participation in this decision, but he found none (personal communication, 1985). Also, perusal of the annual report of the Bureau of Indian Affairs reveals no consultation with Zuni tribal officials. It must be recalled that the Zuni tribe had a series of secular governmental officials initially established by the Spanish, and these offices continued through the Mexican period. The religious leaders appointed a governor *(tapupu)*; then he, after consulting with these religious authorities, selected a lieutenant governor along with *tenientes* from each farming village. These officials were in place at the time construction began, and although the governor was addressed about other matters influencing the welfare of Zuni, such as the smallpox epidemic of 1898–99, no such correspondence concerning the dam has ever materialized. The decision to build a dam at Black Rock apparently was reached quickly, without consultation, at the end of the last century. Congress authorized the dam and irrigation projects on July 22, 1903.

Construction of the dam on Zuni land did not guarantee preferential employment for Zuni men. Workers came from all over the Southwest—Hopi, Laguna, Navajo, Mexican, and Anglo—to run the power equipment used to haul the basalt blocks into place and to provide physical labor. In 1904, only about forty Zuni men were employed every day; they were paid a dollar a day. This level of employment was not adequate incentive for the Zuni to endorse the project with any degree of enthusiasm. One wonders what the Zunis thought about a dam project that experienced cost overruns attributed to them, two serious breaks in its wall, and that irrigated land insufficient even for the agricultural needs of the Zuni population in 1910.

As stated earlier, lands in the Zuni Irrigation Project below the dam were not classified as good for corn agriculture. The Zunis did not farm this land prior to the building of the dam, and the project has never been successful, for a variety of reasons. The irrigation land is tight clay, the least preferred by the Zuni for growing corn. Farming it, even with irrigation water, is tedious, and it supports many more weeds than do loam and sandy soils. The Zuni were willing to grow wheat on such land, but not corn, beans, or squash. Silt displaced the water behind the dam quite rapidly, precipitating a crisis within two decades of completion. The silt was the product of clear-cutting of forest in the off-reservation Zuni Mountains and overgrazing in the eastern watershed of the Zuni River and its tributaries. As a result there is insufficient flow to carry water the length of the canal.

From its inception, the Black Rock Dam had an unfortunate and damaging consequence to Zuni culture, and particularly to a major sacred spring that is now buried beneath the useless silt impounded behind the basalt dam. According to Zuni oral history, Malokyatsiki, Salt Mother, once lived in the spring, *ḳya'nanaḳnana*. Feeling that such proximity to the village was disadvantageous, she left for what is now Zuni Salt Lake. Before leaving the black rocks with the Gods of War, she instructed a youth of the Frog Clan: "In four years I wish your people to come here and put my house in good order." The spring, which measured fifteen by twenty feet, with terraced ledges beneath its surface, was cleaned by women of the Frog Clan according to the wish of Salt Mother and under orders from the Pekwin. Special pottery bowls were made to hold feather offerings for placement on the ledges.

The unfortunate end of this ceremonial practice was that workmen building the dam apparently desecrated the spring. In an informative footnote to the legend of Mawe's flight to the Salt Lake, Stevenson (1904:60) noted: "These sacred objects will soon be scattered, as the secret of burying the vases beneath the water has become known to the men now employed in constructing the Government dam for these Indians. This spring will be in the bed of the great reservoir." In a sad conclusion to his excellent paper about the special importance of water to the indigenous people of the Southwest, Hough independently recorded the final destruction of this sacred spring: "Near Zuni, the engineers who are building a great dam to impound the waters of the Zuni River cleaned out in 1904 an excellent spring, and in the debris found many ceremonial objects, which, unfortunately, were not preserved. This spring, which has an important place in Zuni tradition, wells up through a deposit composed of the remains of Pleistocene animals, and on completion of the dam it will be submerged"(Hough 1906:169). From an ecological perspective, this spring could only be maintained with the proper ceremonies, if its beneficence was to sustain the people.

Reservoirs

Serious problems have limited the agricultural potential of earthen reservoirs on the Zuni reservation. It must be remembered that the Zuni built small dams for water diversion themselves. It was not until 1929 that the government began to modify them by adding more dirt to their tops or by building large replacement dams. These efforts have failed because of silt filling them and displacing water from off-reservation clear-cutting and from on-reservation overgrazing.

Another serious consequence of the Black Rock Dam and other water containment projects on the Zuni reservation was extensive erosion of the farmlands, according to conclusions reached by the Zuni legal team (Ford 1985). With each dam the rivers had to seek a new equilibrium point, and this led to extensive erosion after 1910. A social problem was created when government officials encouraged Zuni to farm the new lands at the expense of their outlying fields, where they had controlled the water and thus prevented erosion for centuries. Now the villages became depopulated and the land lost its traditional water managers. The farming methods described above, once practiced extensively both on and off the present reservation, had been sufficient to limit backcountry erosion, but to be effective they required the presence of numerous farmers. Without the traditional stewards of the land, erosion could and did proceed unabated.

Overgrazing

Overgrazing did not become a serious problem until after 1934, when the Zuni reservation was fenced in. Before that, Zuni ranchers had herded sheep on aboriginal land, south and west of the reservation. Although individual flocks often numbered in excess of two thousand animals, there was ample land for grazing. Once the area was fenced, however, the U.S. government required the Zuni to withdraw from those aboriginal lands unless they owned or leased them, and to move livestock onto reservation land. Immediately the carrying capacity was exceeded in many areas, and individual land allotments were inadequate for the numbers of animals now concentrated there. A resident recalls: "You know, this reservation was not here, a long time ago. Like we said, people who had sheep . . .were grazing way out to the south, west, north, or east. But then when this fence came, they were all crowded in, maybe about fifty to sixty thousand head of sheep. Just like, you know, in a small fence, with a lot of animals in here. So that's when the erosion came about." Eventually, in 1942, stock reduction was employed to limit herd size, but the unexpected administrative mandate almost a decade earlier had already brought about land damage.

The stock concentration program created other problems for the Zuni farmer. Previously, if livestock had damaged a field, the owner of the animal was held accountable and had to pay damages. Some fields had brush fences to discourage animals, and a few had barbed wire, but most farmers relied on custom to keep their fields free from damage. With the encouragement of livestock as an economic base, the rules changed. Now an unfenced field damaged by stock was the farmer's problem. If he could not afford wire, he had to abandon his field. Many tributary fields were vacated because of the uncompensated damage livestock caused.

Left to their traditional means of regulating herd size, grazing, and protecting farms, the Zuni encountered few major problems from livestock. Once the government interceded and usurped control over Zuni affairs, the imposed changes had negative consequences for both the Zuni and the land.

Conclusions

The Zuni case demonstrates that ethnoscientific methods are indispensable for elucidating ecological problems. Without the benefit of folk terms, classifications, and rules for behavior toward natural phenomena, ecological anthropology is nothing. At the same time, it is necessary to have the tools of Western science to determine the impact that a culture's beliefs and behavior actually have on the natural environment. It cannot be assumed that local knowledge is of no consequence for understanding the structure of the physical world. That was the mistake the U.S. government made at Zuni: It ignored generations of experience and native beliefs and created an unintended and extremely costly ecological disaster.

This case never came to court, and many of the issues raised by it were not resolved. Congress intervened and passed—and President Bush signed—the Zuni Land Conservation Act of 1990 (Public Law 101-486). Under its provisions, the Zuni cannot sue the government again over these conservation issues or the loss of natural and cultural resources. The millions of dollars the Zuni were awarded were placed in trust, with the interest to be used by the Zuni for land restoration according to their own methods. Today, the Zuni are once again using traditional techniques of land and resource stewardship under the direction of their own people to restore the land and to keep the trust land from further destruction.

References

Berlin, B. 1992. *Ethnobiological Classification*. Princeton: Princeton University Press.

Bunzel, R. L. 1932. "Zuni Ritual Poetry." *Forty-seventh annual report of the Bureau of American Ethnology (1929–30)* 47:611–837. Washington: United States Government Printing Office.

Conklin, H. C. 1957. *Hanunoo Agriculture: A Report on an Integral System of Shifting Agriculture in the Philippines*. Rome: FAO, United Nations.

Cushing, F. H. 1920. *Zuni Breadstuff*. New York: Museum of the American Indian.

Fewkes, J. W. 1891. "A Few Summer Ceremonials at Zuni Pueblo." *Journal of American Ethnology and Archaeology* 1:1–62.

Ford, R. I. 1985. "*Zuni Land Use and Damage to Trust Land*. Exhibit 7000." Expert testimony submitted to the United States Claims Court as evidence in the case *Zuni Indian Tribe v. United States*, Docket 327-81L.

Fowler, C. 1977. "Ethnoecology." In *Ecological Anthropology*. D. L. Hardesty, ed. New York: John Wiley & Sons.

Frake, C. O. 1962. "Cultural Ecology and Ethnography." *American Anthropologist* 64(1):53–59.

Hardesty, D. L. 1977. *Ecological Anthropology*. New York: John Wiley & Sons.

Hough, W. 1906. "Sacred Springs in the Southwest." *Records of the Past* 5: 165–69.

Johnson, A. W. 1974. "Ethnoecology and Planting Practices in a Swidden Agricultural System." *American Ethnologist* 1:87–104.

Lansing, J. S. 1991. *Priests and Programmers*. Princeton: Princeton University Press.
Nazarea-Sandoval, Virginia D. 1990. "Ethnoagronomy and Ethnoagastronomy: On Indigenous Topology and Use of Biological Resources." Paper presented at the second International Congress of Ethnobiology, Kunming, China.
———. 1991. "Memory Banking of Indigenous Technologies Associated with Traditional Crop Varieties: A Focus on Sweet Potatoes." In *Sweet Potato Cultures of Asia and South Pacific*. R. E. Rhoades and V. Nazarea-Sandoval, eds. Los Baños: User's Perspective with Agricultural Research and Development.
———. 1995. *Local Knowledge and Agricultural Decision Making in the Philippines: Class, Gender, and Resistance*. Ithaca and London: Cornell University Press.
Orlove, B. S. 1980. "Ecological Anthropology." *Annual Review of Anthropology* 9: 235–73.
Rappaport, R. A. 1965. "Aspects of Man's Influence upon Island Ecosystems: Alteration and Control." In *Man's Place in the Island Ecosystem*. F. R. Fosberg, ed. Honolulu: Bishop Museum Press.
———. 1968. *Pigs for Ancestors*. New Haven: Yale University Press.
Stevenson, M. C. 1904. "The Zuni Indians: Their Mythology, Esoteric Fraternities, and Ceremonies." *Twenty-third annual report of the Bureau of American Ethnology 1901–2*. Washington: U.S. Government Printing Office.
Stewart, G. R. 1940. "Conservation in Pueblo Agriculture: II." *Scientific Monthly* 51(October): 329–40.
Sturtevant, W. C. 1964. "Studies in Ethnoscience." In *Transcultural Studies in Cognition*. A. K. Romney and R. G. D'Andrade, eds. *American Anthropologist* 66(3,pt.2): 99–131.
Vayda, A. P., and B. J. McCay. 1975. "New Directions in Ecology and Ecological Anthropology." *Annual Review of Anthropology* 4:293–306.
Vayda, A. P., and R. A. Rappaport. 1968. "Ecology: Cultural and Non-Cultural." In *Introduction to Cultural Anthropology*. J. A. Clifton, ed. Boston: Houghton Mifflin.

PART II

A Parallax Recognized

Refocusing Ethnoecology's Gaze

6

Lenses and Latitudes in Landscapes and Lifescapes

VIRGINIA D. NAZAREA

Ethnoecology deals with human cognition of environmental components such as plants, animals, water, and soils—a classification of its features that guides action within that environment. Our research in Lantapan, Bukidnon, in the Philippines ("The Ethnoecology of the Manupali Watershed," a collaborative research project funded by the Sustainable Agriculture and Natural Resources Management Collaborative Research Support Program) explores the interface between the landscape and the lifescape and the resulting resource management practices of the local population. If a landscape is construed to be "a mosaic of interacting components with both commonalities and uniqueness" (Sustainable Agriculture and Natural Resource Management and Collaborative Research Support Program 1991), a lifescape can be visualized as the superimposition of human intentions, purposes, and viewpoints over environmental features and the resulting patterns of production, consumption, and distribution (fig. 6.1). The landscape, or what's out there, is processed through human perception, cognition, and decision making before a plan or strategy is formulated and an individual or collective action is executed. Incursion into this fuzzy interface is important and challenging because, as Wolf (1984:397) pointed out: "We do not attack reality only with tools and teeth, we also grasp it with the forceps of the mind—and we do so socially, in social interaction and cultural communication with our fellows and our enemies."

In the Manupali watershed in Lantapan, Bukidnon, the landscape has been shaped by waves of migration from neighboring as well as more distant islands of the Philippines, in response to the availability of land, the cooler climate, and the favorable growing conditions. Thus, layers of ethnicity—composed of native Bukidnons or Talaandigs, Dumagats or Lumads from the

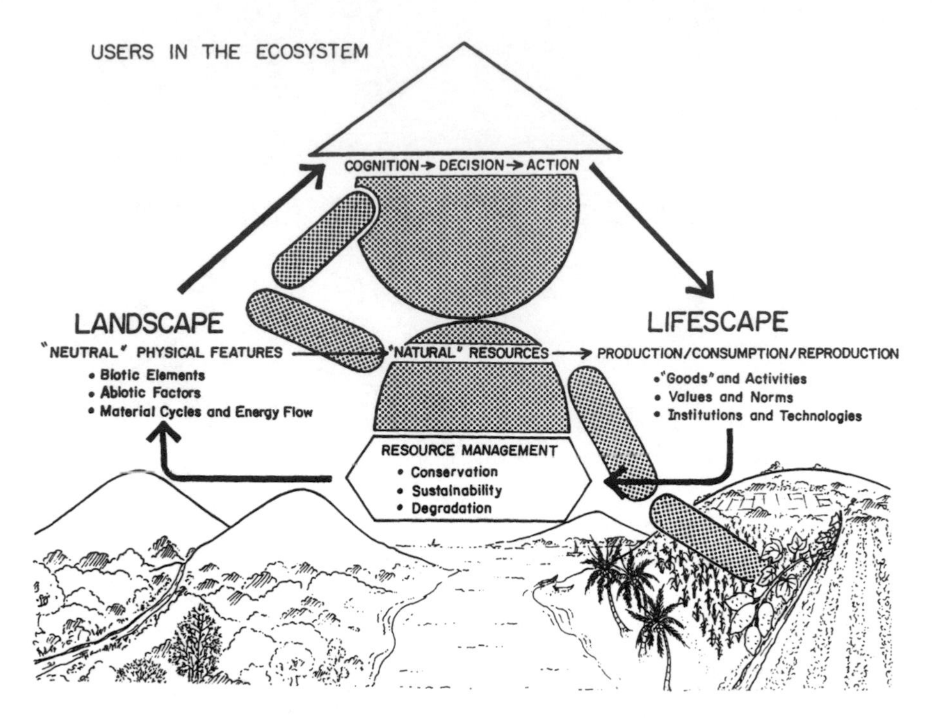

Figure 6.1. Conceptual model for landscape-lifescape integration.

Visayan Islands, and Igorots and Ilocanos from northern Luzon have transferred and reworked environmental knowledge as well as agricultural beliefs and practices in a shared landscape—the Manupali watershed and its surrounding areas. Ethnic boundaries are maintained to the extent that several dialects are spoken and ethnic jokes and commentaries spice up conversations, both private and public, but Visayan is spoken and understood by everyone, and intermarriage is not uncommon. The resulting patchwork of production strategies includes irrigated rice, corn, and sugar cane production in the lowlands, and vegetables, root crops, corn, coffee, and rain-fed rice in the uplands.

I thought it would be intriguing to examine how different categories of people, disaggregated, for example, according to ethnicity, gender, and age, perceive the shared landscape as well as the social, cultural, political, and economic relations that compose the lifescape. Are there, in other words, different lenses that color, magnify, distort, and blur the same set of ostensibly neutral biophysical features for different categories of "viewers" and "actors"? If we understand sustainability to refer to "how people in a particular locale manage resources both in order to maintain themselves on a daily basis and to insure they have what they need as they move from one annual cycle to the next and from one generation to another" (Collins 1991:33), we might ask, How do people arrive at decisions about levels of exploitation along with questions of rightful distribution and access that permit certain strategies and disallow others? In other words, to what extent are landscapes ideological constructions and in what ways do lenses and latitudes color the culturally relevant indicators of sustainability?

Criticizing the commodity focus in agriculture, Ann Ferguson (1994: 541) pointed out that "the socioeconomic and cultural dimensions of sustainability and their relationship to processes of social reproduction are seldom explored . . . [and] links between sociocultural and environmental sustainability are usually overlooked." Social scientists are currently engaged in addressing this oversight. Billie DeWalt (1994:123), for instance, has emphasized that "construction (or, in many cases, reconstruction) of a more sustainable and socially just agriculture has led many individuals to argue that we need to give greater attention to indigenous knowledge systems . . . based on 1) the need to create more appropriate and environmentally friendly technologies, 2) empowering people like farmers to have greater control over their own destinies, and 3) creating technologies that will have more just socioeconomic implications." In short, there is a crucial gap in the sustainability equation that needs to be addressed. This concerns the cognitive or ideological underpinnings that shape the culturally relevant indicators of sustainability which, unfortunately, are impossible to tap using survey questionnaires.

In this paper, I attempt to disaggregate perceptions of the environment based on ethnicity, gender, and age by "recycling," with some modifications, a classical projective technique—the Thematic Apperception Test (TAT). I refer to the product as fragmented ethnoecologies to stress the plurality of understandings reflecting, quite systematically, the standing of actors or decision makers in the internal differentiation of society, a disposition acquired as a product of a position in a multidimensional social sphere (Bourdieu 1987). Elsewhere, Chandra Mohanty (1991) used an even stronger, more vivid phrase, "cartographies of struggle."

The relevance to ethnoecology in particular and ethnoscience in general is the shift in the focus of the debate from whether intellectualist or utilitarian motivations hold sway (Berlin 1992; Ellen 1993) to whether, and in what ways, culture structures perception and operates both as a way of seeing and as a way of not seeing things. This leads us to questions that have been begging to come out from under the rug. To what extent is knowledge shared and to what extent is it systematically patterned? How do different environmental models affect decisions and actions? What drives differential distribution of knowledge, and what is the possibility of breaking free of cognitive bondage? I submit that these questions constitute a watershed problem, not only in the ecological sense in the Manupali landscape but also in the theoretical sense in the landscape of ethnoecology.

Background on Method

The Thematic Apperception Test (TAT) was developed by Henry Murray at Harvard in the 1930s. It consists of a set of cards each showing an ambiguous representation, generally drawn, of one or more human figures. The informant or respondent is asked to tell a story about each card as it is presented, and the resulting account is recorded verbatim. Respondents tend to identify with one or more of the figures, thus revealing some of their own self-concepts and deep wishes in the process of storytelling without much conscious effort. Verbal accounts have traditionally been scored based on Murray's theory of human needs.

When used in comparative, cross-cultural investigations, TATS are modified to make the figures and situations more familiar to the respondents. Alternative scoring methods have been developed for various purposes, such as Gladwin's and Saranson's classical research on Trukese model personality (1953) and Roe's study on personalities of highly successful natural and social scientists (1953). Fine (1955) proposed a system of using TATS and other verbal projective techniques based on "manifest content" of stories, each of which is treated as a distinct unit. The stories, according to Fine's scheme, were scored simply on the presence or absence of previously identified categories. In his particular case, these general categories were feelings, outcomes, and interpersonal relations. Each was further broken down into more specific indicators and states.

For this study, I took photographs of scenes around the Manupali watershed depicting various production strategies, ecologically sound and not- too-sound practices, gradients from monoculture to diverse plots, and human relations with plants and with other humans. These were used as stimuli to elicit cultural conceptions of human-environment interactions and to identify culturally relevant indicators of sustainability and quality of life. After pretesting with two respondents—a male and a female belonging to different ethnic groups—the set of twenty photographs was shown to eighteen informants in each of the three *sitios* (villages) within the research area, divided more or less equally among the three ethnic groups and with approximately equal representation of males and females. An analysis of the results of the dominant themes in informants' stories is presented here. The potentials and limitations of modified TATs as a method of elicitation of culturally relevant indicators of sustainability are also discussed.

Analysis of Modified Thematic Apperception Tests (TATs)

After the stories elicited by modified TATs were collected in Lantapan, we developed a scoring system in the Ethnobiology/Biodiversity Laboratory that focused on culturally relevant indicators of sustainability and quality of life. The system is roughly based on Fine's (1955) scoring scheme. However, the major categories were not divided a priori but instead were arrived at only after a careful reading of several randomly chosen informant responses in an effort to identify the major themes and subthemes. As a consequence, and as can be expected, the major themes differed drastically from Fine's. These were the dominant themes that became apparent to us as we read through the informants' TAT responses:

1. intrinsic values that have no bearing on direct use or the commercialization of natural resources—including comments on the beauty of the environment, the beauty of crops, interconnectedness or interdependence of environmental components, order, and sense of place.
2. concerns about the care and use of the environment that encompass the value of natural resources for direct use by human beings as well as ideas about human responsibility toward these resources—including, for example, comments on the diversity of plants, the cleanliness of the surroundings, the health of the biophysical environment, the usefulness of environmental features (including plants), a sense of stewardship, overexploitation of the environment, and ambivalence regarding the sustainable level of exploitation.
3. production orientations that reveal concerns beyond direct utility toward commercialization—including such subthemes as the health of crops and livestock, industriousness in caring for plants and the environment, the commercialization of production, the procurement of agricultural inputs, luck in productive ventures, and the appreciation of indigenous beliefs and practices.

4. social, economic, and political relations that consist of human relations at the family, community, and governmental levels—including views of family relations, community relations, health, happiness, and the beauty of humans; improvement of livelihood, social stratification, and differential access; government control, authority, or program; and difficulty of life.

After developing the scoring system, we set out to investigate which concerns were most salient to the local people by scoring individual responses to the TATS. Two Ph.D. students at the University of Georgia—Erla Bontoyan from the Department of Statistics and Gabriela Flora from the Department of Anthropology—proceeded to read carefully through each informant's responses to all twenty plates composing the TATS. They scored a theme only when they agreed that a dominant theme had been cited in the informant's story. If there was any difference of opinion between them, the impasse had to be broken by Shankar Talawar, a postdoctoral fellow at the laboratory. Essentially, then, for a theme to be scored, there had to be agreement between a Filipino statistics student and an American anthropology student. Further, any lack of consensus had to be resolved by an Indian postdoctoral fellow with a background in agricultural extension. I did not participate in the scoring.

After the dominant themes were scored for each informant, data were analyzed using the Statistical Analysis System (SAS 1995). Means and standard deviations of total scores of the themes were calculated using PROC MEANS. To determine statistical differences in score means for different themes across ethnicity, gender, and age, analysis of variance through the General Linear Procedure Model (PROC GLM) was conducted. A significance level of 0.10 was used to test statistical significance. Significant differences in means exist if the probability (p) of observing a more extreme value that the test statistics (f) is less than 0.10. The results are discussed in the following section.

Points of View

According to Roger Keesing (1987:161), "Views of culture as collective phenomena need to be qualified by a view of knowledge as distributed and controlled. . . . We need to ask who creates and defines cultural meanings and for what ends." In the same vein, Bob Scholte (1984:540) pointed out that "one cannot merely define men and women in terms of webs of signification they themselves spin since . . . few do the actual spinning while the majority is simply caught." How is knowledge about agriculture and the environment distributed among different actors in the Manupali watershed and how does this distribution affect their perspectives on a shared and contested "stage"? A careful examination of the results of the Thematic Apperception Test indicates some dominant themes and their distribution across ethnicity, gender, and age.

Men, in general, focused a great deal of attention on political and economic situations and the constraints and malleability inherent in these situations. They made a significant number of very perceptive comments on social stratification, inequities of distribution, and access to resources and rewards. They also demonstrated considerable knowledge pertaining to cash crops and market forces. By comparison, women focused more on family concerns, particularly child bearing and child rearing. They also paid significant attention to community relations, particularly to the maintenance of an "image," a point I will return to later. Moreover, women generally possessed detailed knowledge of traditional practices and food preparations as well as the medicinal value of plants.

The native Bukidnons or Talaandigs, as well as the Visayan and Ilocano/Igorot migrants, frequently alluded to the beauty of the surroundings as depicted in the photographs. "Beautiful" was also used as an adjective for people, for fruits, for vegetables, and for pleasing scenery. It was used interchangeably with the phrase "nice to look at." Visayans or Dumagats frequently referred to the payoffs of taking risks in agricultural ventures, citing "hitting the jackpot" as a motivation for certain decisions or as an explanation for the economic success of their neighbors. On the other hand, Ilocanos and Igorots concentrated more on the state of things as reflecting process or conditions, repeatedly pointing out how clean and "healthy" gardens, or dilapidated houses, flowerless front yards, and neglected fields, reflected either the industriousness or the laziness of their owners.

Going back to our analysis using the scoring systems we developed in the laboratory, figure 6.2 shows that based on percentage of total number of responses for the entire population, the ten most dominant themes, ranked in order of decreasing importance, are: (1) usefulness, (2) commercialization, (3) beauty of the environment, (4) appreciation of indigenous knowledge, (5) diversity of plants, (6) interconnectedness, (7) social stratification, (8) family relations and, equally ranked, government authority, (9) health of crops and, equally ranked, health of humans, and (10) industriousness. The two most highly ranked themes demonstrate a kind of symmetry with operationally defined "indicators" in the sense that both usefulness and commercialization have a direct bearing on productivity and can be reflected in increased Gross

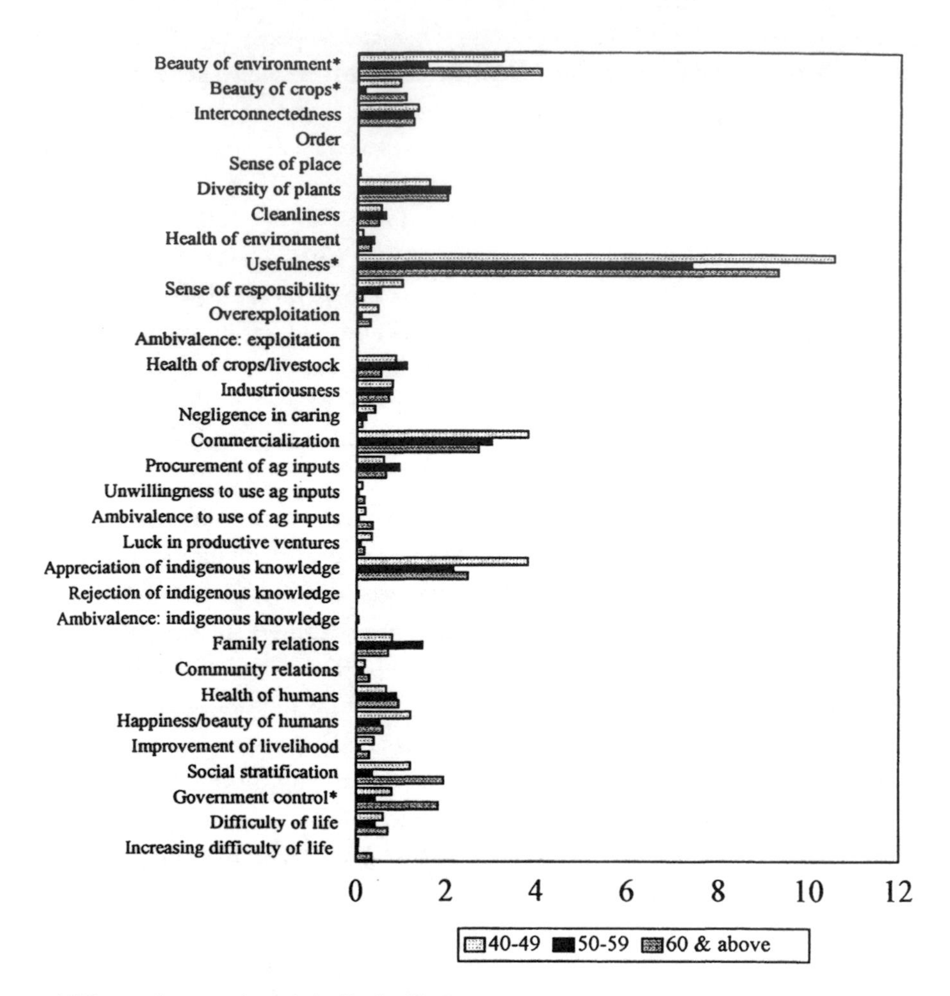

Figure 6.2. Comparison of means of dominant themes for general population.

National Product (GNP). However, the next four, as well as the ninth and tenth, themes are more related to quality of the environment and quality of life and do not translate to any direct—particularly economic—measurement of development. On the other hand, the salience of social stratification, family relations, and governmental authority reveals that the informants are very much

Table 6.1. Means of themes that are significantly different, by ethnicity.

	MEAN (STANDARD DEVIATION)			
Theme	Dumagat	Ilocano	Talaandig	F
Beauty of crops	0.72 (1.07)	1.06 (1.35)	0.20 (0.56)	2.63*
Usefulness	9.61 (2.89)	7.50 (3.43)	9.87 (3.66)	2.64*
Luck in productive ventures	0.11 (0.32)	0.39 (0.61)	0.07 (0.26)	2.80*
Social stratification	0.83 (1.10)	0.50 (0.86)	2.27 (3.79)	2.88*
Increasing difficulty of life	0.11 (0.32)	0.00 (0.00)	0.40 (0.74)	3.50**

*$p < 0.10$; **$p < 0.05$.

aware that they are operating in a system of uneven power relations and are aware of the latitudes they have by virtue of their position.

Disaggregating the dominant themes by ethnicity (table 6.1), we can see some statistically significant differences in the relevance of certain themes to different ethnic groups. The Talaandigs pay less attention to "luck" in productive ventures, as well as beauty of crops, and more attention to direct usefulness of environmental features. This is congruent with the greater degree of subsistence orientation among the Talaandigs and the greater degree of commercialization of production systems of the Ilocanos and the Dumagats.

The following comments abstracted from their stories are particularly revealing:

1. Talaandig female: "What is this? A tree or a flower? It is important for us old people to plant trees and flowering plants. I plant them around the house so that if someone gets sick, it is easy to pick some medicine. Our grandmothers during our time insisted that we plant flowering plants around our house so that bad spirits will be distracted by beautiful flowers and not enter our house.
2. Dumagat male: In this last picture, we can see a garden planted with cabbage. It is almost ready for harvest because they are already big. These crops are very useful in life. For those who have capital, it is like hitting the jackpot when the price goes up.

Further, the three ethnic groups differ most significantly in terms of their perception of increasing difficulty of life and, secondarily, of social stratifica-

Table 6.2. Means of themes that are significantly different, by sex.

Theme	MEAN (STANDARD DEVIATION)		
	Male	Female	F
Sense of responsibility	0.16 (0.37)	0.88 (1.92)	3.42*
Negligence in caring for environment	0.12 (0.33)	0.35 (0.56)	3.03*
Commercialization	2.08 (2.18)	4.15 (3.35)	6.80**
Appreciation of indigenous knowledge	3.64 (3.50)	1.88 (2.57)	4.20**

*$p < 0.10$; **$p < 0.05$.

tion. This result is noteworthy because it indicates that the indigenous population of the Manupali watershed, the Talaandigs, are the most cognizant of, and possibly the most vulnerable to, the weight of increasing commercialization, stratification, and pressure on resources. As one Talaandig informant noted in his story about one of the pictures: "In the second picture, there is a home constructed out of split bamboo. Most people nowadays have houses made of bamboo because only a few of them can get wood from trees anymore. Forest guards will arrest them. We should be allowed to get wood for our houses because we do not sell them anyway. . . . It is needed, it is justified. However, what the government probably wants is for us poor people to live like rats and just find ourselves a hole and live there."

Comparing male and female informants (table 6.2), we see that commercialization is more salient for males than for females, while the reverse is true for appreciation of indigenous knowledge. Moreover, men emphasize sense of responsibility, which can be interpreted to imply control, and pay more attention to negligence in caring for agricultural crops and natural resources. Take for example this comment from an Igorot/Ilocano man: "Here, this is all good because there are yams. These can be eaten by the people, especially those who planted them. So the owner did not waste his time in caring for these. And then there are also trees in the surroundings. Trees can give fresh air to the people. This is all I can say." By comparison, this comment from an Igorot/Ilocano woman indicates a focus on direct use and indigenous beliefs and practices rather than commercialization and control: "This is sugar cane. Ah, this picture is a sugar cane field used to make sugar. The reason why the Philippines is planting sugar cane is for export. But actually it can also be eaten because it has a lot of juice to quench the thirst of those who are thirsty. It can also be cure for cough and used to clean teeth."

Although not statistically significant at the level tested, stratification and government authority appear to be more salient for males, while quality of life indicators such as family relations and health are somewhat more important for females. The gender role differentiation relative to cognitive orientation therefore occurs along the lines of cash vs. subsistence economy and public vs. private sphere, indicating a more outward orientation for males and inward orientation for females. Another implication of these findings, which has relevance to the literature on women and plant genetic resources conservation, is that women, by virtue of their being in charge of ensuring the day-to-day subsistence of the household as well as their responsibility for homegardens (as opposed to cash crops), generally are more promising partners in situ conservation, or local conservation through use, of culturally significant plants (Nazarea 1998; Sachs 1996; Shiva 1993).

In differentiating informants according to age, one thing that stands out is the fact that those in their fifties depart quite dramatically from those in their forties and those in their sixties or older with respect to salience of most of the themes. Of all the age groups, the fifties category paid the least attention to beauty of the environment, beauty of crops, sense of place, usefulness of environmental features, and overexploitation. They expressed the least unwillingness (or, by inference, the greatest willingness) to use agricultural inputs and the least ambivalence toward agricultural inputs. Informants in their fifties also voiced the least concern for luck in productive ventures, improvement of livelihood, social stratification, government authority, and difficulty of life. On the other hand, compared with all the other age groups, this group paid the most attention to interconnectedness, diversity, cleanliness, the health of the environment, the health of crops, industriousness, procurement of agricultural inputs, rejection of indigenous knowledge, and family relations.

The difference among age groups is most significant for beauty of crops and usefulness and, secondarily, for beauty of environment and governmental control (table 6.3). Holding ethnicity and gender constant to facilitate comparison based on age, the difference in perspective among farmers in their sixties, fifties, and forties is evident in the following TAT stories:

> Dumagat male, sixties: I can see a drum which is a container for tomatoes and there are grasses for the animals to eat. There are also

Table 6.3. Means of themes that are significantly different, by age.

Theme	MEAN (STANDARD DEVIATION)			
	40–49	50–59	60 and Older	F
Beauty of environment	3.20 (2.62)	1.53 (1.64)	4.06 (4.93)	2.69*
Beauty of crops	0.93 (1.22)	0.16 (0.50)	1.06 (1.30)	3.93**
Usefulness	10.53 (2.92)	7.37 (3.35)	9.29 (3.35)	4.17**
Government control	0.80 (1.21)	0.42 (0.77)	1.82 (2.90)	2.68*

*$p < 0.10$; **$p < 0.05$.

different kinds of tomatoes; the red and the white ones as well as the yellow ones. These are good "appetizers" for our meals.

Dumagat male, fifties: These are beans. Are these beans? This is used as a vegetable and you can get a lot of money from them everywhere in the Philippines. What are these, tomatoes? There is also a lot of money to be made from these, so there are many people planting them. But this is not a project here because we do not have fertilizers. We cannot grow them here. It will just be damaged here. They call it "wilt."

Dumagat male, forties: This is corn, which is important like rice. This is commonly used by people. I cannot examine clearly the kind of seedlings used here but I think these are useful to the farmers. If we do not have corn, there is nothing to eat, and it is important because we can also sell it. The corn meal is consumed by the people and the corn bran is used to feed pigs, goats, and cows.

It may be recalled that farmers who are presently in their fifties were coming of age (in their early to midtwenties) during the heyday of the Green Revolution in the Philippines, with its emphasis on high-yielding varieties and commercial, high-input agriculture. Those currently in their sixties were then in their mid to late thirties and perhaps less impressionable, while farmers now in their forties were then perhaps still too young to care. While the results support the lamentable contention that the Green Revolution ideology fostered a sense of control over, rather than harmony with, the environment, they also demonstrate that the younger generation is reverting to an appreciation of the values of a generation once removed.

Conclusions

Bourdieu (1987) defines a point of view as "a perspective, a partial subjective vision . . . but . . . at the same time a view, a perspective, taken from a point, from a determinate position in an objective social space." He goes on to say:

> Any theory on the social universe must include the representation that agents have of the social world and, more precisely, the contribution they make to the construction of the vision of that world and, consequently, to the very construction of that world. It must take into account the symbolic work of fabrication of groups. . . . It is through this endless work of representation that agents try to impose their vision of the world or the vision of their own position in that world, and to define their social identity. By the same token, it must be recognized that agents involved in this struggle are very unequally armed in the fight to impose their truth, and have very different, even opposite, aims.

The results of the modified TATS demonstrate that disposition is very much a product of position. Although certain perceptions and priorities are shared, different ethnic, gender, and age groups view the landscape in a manner characteristic of their social niche and particular histories. Yet, granting that people, or agents, are "very unequally armed," the individual stories underscore the depth of consciousness regarding the value of natural resources as well as the skewed distribution of, and unequal access to, resources at every rung of the social hierarchy. Hegemony may be robust, but it is never very thorough. Whether they are referring to how the house of the parents of the forest ranger was constructed of valuable hardwood that no one is supposed to cut, or pointing to alternatives such as vegetable growing that are available only to those with considerable capital, people belonging to different ethnic and gender categories in Lantapan puncture or poke fun at the official "theodicies" or the dominant version of reality at every turn, while at the same time pragmatically recognizing the bounded latitudes in which they must operate. Based on a strong sense of place, they express, again and again, an urgent need for self-determination in the management and conservation of natural resources.

Another dominant theme, this time one that cuts across ethnic and gender lines, is the salience of beauty of the surroundings; of fields and produce that are "nice to look at," of houses and gardens that need to be "presentable," and of well-tended crops revealing the care invested by their owners. I take this to be reason for optimism following the argument from deep ecology that affective motivation or emotional identification with nature, as expressed in awe regarding its beauty, is a force that is much stronger than logical, objective justifications for environmental conservation. Beauty emphasizes quality, a comforting "rightness" about one's near and far environment that moves one to care for, conserve, and protect.

In terms of method, I believe that Thematic Apperception Tests complemented by oral history analysis and cognitive mapping can provide us with a window to the patterning of local perceptions regarding the environment and a way to apprehend the culturally relevant indicators of sustainability. According to Kaplan and Saccuzzo (1989), the TAT is still one of the most important techniques used for clinical tests and research. For our purposes, some improvements can be made in the direction of making the plates or pictures more ambiguous, to leave more room for interpretation and to introduce a time depth or change dimension by including "before" and "after" shots of natural resource clusters obtained through archival research or a more prolonged involvement in the community. We are currently developing sets of three plates that depict processes pertaining to biodiversity, soil erosion, and water quality for use as TAT stimuli.

The search and the effort, however tentative, are justified by the need to understand how local perceptions affect decisions and actions in the landscape, as opposed to how local resource management—the interface between landscape and lifescape—can be characterized, evaluated, and intervened in, based on external, operational parameters. What I think the modified TATS tapped into are the ambiguities regarding what constitutes a good life and a good environment. If the quality of life and sustainability of the landscape are regarded as matters of fact, then the only worthwhile approach to their investigation would be to measure relevant variables by increments and establish some sort of critical threshold. Moreover, if the society or culture is homogenous, then one supposedly omniscient observer/informant is all we need to factor in the human dimension. What we have seen, however, is that divergence and patterning—not uniformity and universality—characterize human cognition of the environment that presumably guides their actions in it.

What constitutes the "landscape" that social and natural scientists are trying so assiduously to assess, protect, and restore? In a scenario that gets "curiouser and curiouser," we realize that in every human-dominated system we are dealing not just with the neutral biophysical landscape but also with the very idea of the landscape; in fact, not just with the idea but with a patterned, nested multitude of ideas. It is somewhat humbling to note that there are differing points of view in implicit or explicit confrontation, and ours is only one among them. In a very real sense, what we are dealing with in every aspect of resource management is people's sense of place—the lenses through which they construct the environment and estimate their latitudes of choice and opportunities for challenge and refutation.

References

Berlin, Brent. 1992. *Ethnobiological Classification: Principles of Categorization of Plants and Animals in Traditional Societies.* Princeton: Princeton University Press.

Bourdieu, Pierre. 1987. "What Makes a Social Class?: On the Theoretical and Practical Existence of Groups." *Berkeley Journal of Sociology: A Critical Review* 31: 1–18.

Collins, Jane. 1991. "Women and Environment: Social Reproduction and Sustainable Development." Pp. 33–58 in *The Women and International Development Annual,* vol. 2. Rita Gallin and Anne Ferguson, eds. Boulder: Westview Press.

DeWalt, Billie R. 1994. "Using Indigenous Knowledge to Improve Agriculture and Natural Resource Management." *Human Organization* 53(2):123–32.

Ellen, Roy 1993. *The Cultural Relations of Classification: An Analysis of Nualu Animal Categories from Central Seram.* Cambridge: Cambridge University Press.

Ferguson, Anne E. 1994. "Gendered Science: A Critique of Agricultural Development." *American Anthropologist* 96(3): 540–52.

Fine, R. 1955. "Manual for a Scoring Scheme for the TAT and Other Verbal Projective Techniques." *Journal of Projective Techniques* 19:306–9.

Gladwin, Thomas, and Seymour Saranson. 1953. *Truk: Man in Paradise.* Viking Fund Publications in Anthropology no. 20. Chicago: University of Chicago Press.

Hunn, Eugene. 1989 "Ethnoecology: The Relevance of Cognitive Anthropology for Human Ecology." In *The Relevance of Culture.* M. Freilich, ed. New York: Bergin and Garvey.

Kaplan, Robert M., and Dennis P. Saccuzzo. 1989. *Psychological Testing: Principles, Applications, Issues.* Pacific Grove, California: Brooks/Cole Publishing Company.

Keesing, Roger M. 1987. "Models: Folk and Cultural: Paradigm Regained?" In *Cultural Models in Language and Thought.* Cambridge: Cambridge University Press.

Mohanty, Chandra. 1991. "Cartographies of Struggle." In *Third World Women and the Politics of Feminism*. Bloomington: Indiana University Press.

Murray, Henry A., et al. 1938. *Explorations in Personality.* New York: Oxford University Press.

Nazarea, Virginia. 1998. *Cultural Memory and Biodiversity.* Tucson: University of Arizona Press.

Rappaport, Roy. 1979. *Ecology, Meaning and Religion.* Richmond: North Atlantic Books.

Roe, Anne. 1953. "A Psychological Study of Eminent Psychologists and Anthropologists and a Comparison with Biological and Physical Sciences." *Psychological Monograph*. General and Applied American Psychological Association.

Sachs, Carolyn. 1996. *Gendered Fields: Rural Women, Agriculture and Environment.* Boulder: Westview Press.

Scholte, Bob. 1984. "On Geertz's Interpretative Theoretical Program." *Current Anthropology* 25:540–42.

Shiva, Vandana. 1993. "Women's Indigenous Knowledge and Biodiversity Conservation." In *Ecofeminism*. M. Meis and V. Shiva, eds. Halifax, N.S.: Fernwood Publications.

"Sustainable Agriculture and Natural Resource Management Collaborative Research Support Project." 1991. A Landscape Approach to Sustainability in the Tropics. Concept Paper/Project Proposal. Athens: University of Georgia.

Wolf, Eric. 1984. "Culture: Panacea or Problem?" *American Antiquity* 49(2): 393–400.

Cultural Landscapes and Biodiversity

The Ethnoecology of an Upper Río Grande Watershed Commons

DEVON G. PEÑA

The Upper Río Grande watershed is a thirty-four-thousand-square-mile area stretching from the Rocky Mountains in southern Colorado and northern New Mexico to the Juarez Valley, across the border from El Paso, Texas (Hay 1963:491). The northernmost third of the watershed encompasses a seven-county area in northern New Mexico and southern Colorado with a predominantly Chicano population of Spanish-Mexican origin.[1] This area is the principal headwaters bioregion of the Upper Río Grande but includes important tributaries of the Arkansas watershed. It consists of a series of high-altitude valleys that drain forested, snow-covered mountains.[2] At the end of the sixteenth century, coming north from Mexico, a diverse people began to settle in the intermontane valleys of the watershed.[3] The settlers established agropastoral villages that have been widely praised as examples of sustainable human adaptation to high-altitude, arid-land environments.[4] At the heart of these farm and ranch communities is the watershed commons. The high mountain peaks provide water, timber, pasture, medicinal plants, and wildlife for use in common by the villagers.

Watersheds have traditionally defined the boundaries of self-governing communities in the Upper Río Grande. This invention of political jurisdiction as derivative from a type of hydrographic unit was probably first described by John Wesley Powell in 1890. Writing on the possibilities for sustainable human settlement in the arid-land environments of the intermountain West, Powell observed:

> The people of the Southwest came originally, by way of Mexico, from Spain, where irrigation and the institutions necessary for its control had been developed from high antiquity, and these people well understood that their institutions must be adapted to their industries, and so they organized their settlements as pueblos, or "irrigating municipalities," by which the lands were held in severalty while the tenure of the waters and works was communal or municipal. . . . [The goal of this irrigation tradition was] to establish local self-government by hydrographic basins. (Powell 1890:112–14; see also Worster 1994:1–28)

The organization of acequia-based farms and ranches of the Chicano upland villages of northern New Mexico and southern Colorado is an important example of this watershed commonwealth form of self-governance. The association of acequia members *(parciantes)* is a community of irrigators with shared responsibility in the care, maintenance, and use of the ditch networks. The irrigating municipality is a deeply rooted tradition for effective, local self-management of water and land. The ultimate responsibility of the acequia associations is the management of water rights and stewardship of the watershed commons. This tradition of local self-governance is only now being recognized as a viable alternative in the debate over the future of the commons in the intermountain West (see Wilkinson 1988, 1992). Moreover, acequias are themselves innovations on the rhythms and patterns of the watershed, a type of disturbance ecology that, like beaver works, increases biodiversity by creating wildlife habitat and movement corridors.[5]

Ethnoecology of the Culebra Microbasin

The San Luis Valley in southern Colorado is a high-altitude, cold-desert environment. The valley is topographically and climatically similar to the high steppes of central Asia. The valley has an average elevation of eight thousand feet above sea level and is surrounded by the fourteen-thousand-foot peaks of the Sangre de Cristo Mountains to the east and the San Juan Mountains to the west. The bioregion receives very little rain (with average annual precipitation of seven to eight inches). However, the high mountain peaks on average receive more than one hundred inches of snow during the long seven-month winter season. The moisture from the snow pack is what makes agriculture possible in the valley. The valley is the northernmost headwaters basin of the Río Grande, collecting stream flow from some fifty tributary creeks of the river that originate in the high peaks. One of these tributaries is the Río Culebra, with headwaters in the Culebra Range, the southernmost extension of the Sangre de Cristo Mountains in Colorado.[6]

The Culebra Microbasin

The Culebra microbasin is home to some of the oldest agricultural communities in the state of Colorado. The Chicano villages of the Culebra were settled between 1850 and 1860 by *pobladores* (village colonists) invited by the heirs of the Sangre de Cristo land grant (issued in 1844).[7] The microbasin includes nearly every major life zone in North America, from alpine tundra above timberline (at twelve thousand feet and above) to Upper Sonoran cold desert (at eight thousand feet and below). Montane and subalpine fir forests located at nine to twelve thousand feet are the heart of the watershed. The forest canopy protects the winter snow pack. During the spring and summer the gradual melting of the snow is the primary source of water for irrigation in the microbasin.

Acequias

The irrigation system in the Culebra microbasin is based on the acequia, or gravity ditch system. This irrigation tradition has independent roots in three continents: Africa, Europe, and North America (Peña 1993, 1998). The term acequia derives from the Arabic word *as-Saquiya*, which means the "water bearer" or "water carrier." Acequia irrigation systems are renowned around the world as culturally and ecologically sustainable technologies. They are notable for: (1) a renewable use of water that maintains the equilibrium of the local hydrological cycle through aquifer recharge and return to in-stream flows; (2) a renewable use of energy that relies on the force of gravity to move water; (3) a network of earthen-work ditches that increases biodiversity by creating wetlands and woodlands that serve as wildlife habitats and biological corridors; and, overall, (4) their contribution to the control of soil erosion and maintenance of water quality. The collective community management of the acequia ditches provides a cultural foundation and an institutional tradition for local self-governance and the reproduction of conservation ethics from one generation to the next (see Peña 1993, 1998; Peña and Martínez 1998; Rivera 1997).

To work effectively, acequias rely on the gradual melting of the winter snow pack in the mountains. Any disturbance of the watershed ecology can result in serious problems for acequias. For example, deforestation can lead to excessive sediment loading in the ditches, arroyo cutting, flooding, or lack

Table 7.1. Original acequias, in the Culebra microbasin.

Priority[a]	Ditch	Construction Date
1	San Luis Peoples	April 1852
2	San Pedro	April 1852
3	Acequia Madre	1853
4	Montez	August 1853
5	Vallejos	March 1854
6	Manzanares	April 1854
7	Acequiacita	June 1855
8	San Acacio	April 1856
9	Madriles	April 1856
10	Chalifu	April 1857
11	Cerro	November 1857
12	Francisco Sanchez	March 1858
13	Mestas	May 1858
14	San Francisco	May 1860
15	Trujillo	May 1861
16	Little Rock	1873
17	Garcia	1873
18	Torcido	May 1874
19	Abudo Martin	May 1874
20	Guadalupe Vigil	March 1880
21	Jack J. Maes	March 1881
22	Antonio Pando	April 1881
23	Guadalupe Sanchez	November 1882

[a]As defined by construction date.

of sufficient water during the irrigation season (Costilla County Conservancy District 1993; Curry 1995; Jones and Grant 1996).

In the Culebra microbasin, there are twenty-three historical acequias represented by *mayordomos* (ditchriders) for each of the ditches.[8] These acequias hold the oldest water rights in Colorado under the doctrine of prior appropriation (table 7.1). For example, the San Luis Peoples Ditch, which has the first priority, was constructed in 1852, decreed in 1862, and adjudicated in 1889. The acequias are collectively organized under the umbrella of the Costilla County Conservancy District (CCCD), which was established in 1976. The CCCD has played a major, largely unheralded role in Colorado environmental politics. During the 1970s, the conservancy district led the opposition in successfully opposing a plan by the San Marcos Pipeline Company to mine the local groundwater aquifer to operate a coal slurry line. Repeatedly since the 1980s, the CCCD has been a major force lobbying against the reclassification of farmlands by the Colorado state legislature. These legislative initiatives would change the tax structure so that small farm properties (for example, with fewer than twenty acres or less than $5,000 in annual sales) would lose their status as agricultural land. These initiatives have been thrice opposed by the CCCD because the proposed reclassification would undermine the ability of Chicano smallholders to continue the sustainable tradition of subsistence agropastoralism in the San Luis Valley.

In the late 1980s and early 1990s, the acequias, through the CCCD, led the opposition to the Battle Mountain Gold (BMG) strip mine and cyanide leach mill (see Peña and Gallegos 1993). As a result of the BMG struggle, the CCCD played a major role in a 1993 campaign to reform the Colorado Mined Land Reclamation Act (MLRA). Most recently, the CCCD has played a critical role in the establishment of La Sierra Foundation of San Luis, a community-based organization seeking the return of the Culebra Mountain Tract through a national fund-raising campaign for a community land trust (see Peña 1995a; Peña and Valdéz Mondragon 1997).

The Culebra Mountain Tract

The Sangre de Cristo land grant originally encompassed approximately one million acres. But the traditional commons of the Culebra bottomland villages consists of a 77,754-acre tract that locals know as *la sierra* (Mountain Tract). The Culebra Mountain Tract includes one peak of over fourteen thousand feet in elevation and eight of over thirteen thousand. Until 1995 the area was relatively undisturbed and roadless; there is a two-thousand-acre clear-cut in the southwestern corner of the tract.[9] According to Webb (1983) and Reynolds (1990), the Mountain Tract is historically habitat to nesting pairs of the endangered Mexican spotted-owl *(Strix occidentalis)*, and its creeks are stocked with the native Río Grande cutthroat trout (a rare and threatened species). Most of the Culebra Mountain Tract consists of montane and subalpine conifer forests with a mix of ponderosa, Douglas fir, and spruce. The higher elevations are characterized by alpine tundra, krummholz, and windswept rock lands that are under snow eight to ten months out of the year. Wet montane meadows and marshlands, aspen groves, piñon-juniper woodlands, riparian cottonwood and willow stands, and semidesert sagebrush prairies complete the variety of plant communities in the Mountain Tract.

Headwaters and Agroecology

An important feature of the ethnoecology of this microbasin is the relationship between the alpine and montane headwaters of the Culebra and the farms and ranches located below in the riparian bottomlands. Most local people strongly support the protection of wildlife and its habitat (Peña et al. 1993). Local farmers and ranchers are particularly strong in their support of wildlife conservation through habitat protection because they recognize that the conditions optimizing wildlife habitat also help maintain watershed integrity and water quality (Costilla County Conservancy District 1993). The processes that destroy wildlife habitat and disrupt the watershed are seen to affect farming and ranching negatively. These farms and ranches are notable for their reliance on acequias, use of perennial polycultures, preference for rare native landraces (regionally-adapted family heirloom crop varieties), and the clustering of wildlife habitats and farming landscapes. These farms and ranches are sustainable agroecosystems. A unique cultural-watershed landscape is endangered by industrial capitalist development and extractive activities affecting the ecosystem (Costilla County Conservancy District 1993; Peña and Martínez 1998).

Damage to the watershed presents a definite threat to the ecological basis of these farms and ranches. For example, logging operations destroy wildlife habitat and reduce biodiversity. And such activities also create soil erosion and channel aggradation, diminish water quality, and cause problems with sedimentation for downstream acequias. Deforestation creates flood-control problems with the potential to irreversibly damage the acequias. Deforestation also accelerates the rate at which snow pack melts into stream flow.[10] Too much water comes down too fast at the wrong time. The entire agro-hydrological cycle is thrown off balance.

In the case of the Culebra watershed, limited storage rights for the acequias and overextended storage capacity in structurally unsound reservoirs create the conditions for a hydrological crisis. Farmers and ranchers would not be able to manage the runoff and, lacking storage rights, most of the water for acequias would be lost to in-stream flows before the end of the irrigation season. The lack of sufficient water during the three- to four-month irrigation season would destroy the basis for sustainable agriculture in the microbasin.[11] A long-standing local struggle to restore communal ownership and use of the Culebra Mountain Tract stems from a desire by the irrigating community to prevent this sort of catastrophic damage.[12]

Chicano Agroecosystems: An Ideal-Type

Agroecology provides an interdisciplinary framework for the study of farming communities in environmental and sociohistorical contexts.[13] Agroecology begins with an elegant and seemingly paradoxical premise: Agriculture is, above all else, a human artifact; yet the farming system does not end at the edge of the field. The primary tenet of agroecology is that the farm is itself an ecosystem and part of a larger ecosystem (it is located within a broader bioregional context). Proceeding from the basic recognition of the ecological context of agriculture, this research tradition emphasizes four foundational principles: Agroecology (1) recognizes sense of place as a factor in the coevolution of culture and nature and in the adaptation of agroecosystems to the physical and biological nuances of localities (ontological dimension); (2) values the preservation of local knowledge over the imposition of universal mechanistic knowledge and recognizes the sustainability of traditional agroecosystems (epistemological dimension); (3) privileges the production strategies of traditional polycultures over modern monocultures as a way to correct inequities in agricultural research and extension services (ethical dimension); and (4) empowers farmers by favoring self-management of the natural conditions of production and promoting local control of political economic institutions (policy dimension). (For further discussion, see Altieri et al. 1987.)

Agroecological approaches have not been used in the study of Chicano farming systems. And yet, Chicano agriculture provides a living laboratory for the study of the interactions between cultural, social, economic, political, and ecological systems in a context characterized by limited resources and relatively low levels of mechanized technology. Our preference for the agroecological approach is based on our concern for understanding these practices in a more holistic manner. We also want to endorse an ethically grounded political perspective that supports local initiatives for land reform and democratization of impinging market and state institutions. Given current debates over the future of agricultural policy in the rural intermountain West, the nature of alternative and sustainable models must be made more salient. We must redefine the terms of this debate by outlining a comprehensive and interdisciplinary perspective of Chicano agricultural systems and studying their continuing evolution in contemporary practices.

The sustainability and dynamic character of Chicano agropastoralism is an intriguing possibility, both as a historical legacy and a viable futur · option.

But Chicano agriculture, as a set of living cultural ecological practices, has until now remained relatively unstudied at the level of specific historical research sites.[14] The remainder of this chapter is the first in a series of reports focusing on multigenerational Chicano family farms and ranches that we have designated as historical research sites for an ongoing, long-term study of the cultural and environmental history of the Greater Río Grande watershed.[15] I chose these farms and ranches because they have remained in the same families for five or more generations and continue to be operated as profitable commercial agricultural enterprises.

Chicano agroecosystems in the Culebra microbasin are characterized by several prominent features that are hallmarks of sustainable and regenerative agriculture: (1) a riparian long-lot cultural geography characterized by multiple life zones and ecotones, (2) the use of acequia irrigation systems, (3) the clustering of wildlife habitats and farming landscapes, (4) a tradition of local and regional landraces, (5) the use of natural pest and weed controls with beneficial effects for soil fertility and erosion control, (6) the simultaneous production of several kinds of crops and livestock and an integrated approach to soil conservation and range management, (7) a preference for polycultures and rotational intercropping, (8) the adoption of new soil, pasture, and water conservation practices, (9) a low level of mechanization and a preference for human and animal power, (10) the increasingly common practice of restoration ecology, (11) a tendency toward autarkic prosumption (that is, the production of goods for home and local use and exchange), (12) the maintenance of access to traditional common lands, and (13) an increasingly self-organized and complex set of relationships with a variety of market and governmental institutions.

I present these features—discussed more fully in subsequent sections—as characteristics of an ideal-type. Note that I am not calling this "traditional" agropastoralism. My point of view is that many changes have occurred in Chicano agriculture and that these thirteen features embrace both traditional and more modern practices. Nor am I suggesting that all Chicano agropastoralists are engaged in these practices. Many are not, but these features are prominent enough in most Upper Río Grande microbasins (both historically and contemporaneously) to warrant their inclusion in an ideal- type model. Where possible I have sought to compare and contrast Chicano agroecosystems with mechanized agroindustrial monocultures in order to highlight the sustainability of the agropastoral model.

Riparian Long-Lot Cultural Geography

Agropastoralism in this bioregion depends on a unique and endangered cultural landscape known as the riparian long-lot (fig. 7.1). After passage of the Land Ordinance of 1785, the United States established a national land survey program based on the township-and-range system. According to Donald Worster, this system "divided the country from the Appalachian Mountains to the Pacific Coast into a rigid grid of square parcels one mile on a side, subdivided into quarter sections of 160 acres" (Worster 1994:12). The square-grid system is incompatible with the topographical features and hydrographic boundaries of ecosystems in the intermountain West; it is inconsistent with the lay of the land, water, and native human communities. Anglo-Americans, coming from the East to settle in this region, adopted the square-grid topography of the 1785 land ordinance. This land-use pattern homogenized natural and cultural landscapes by requiring the removal of woodlands, forests, wetlands and other natural and cultural features that were considered obstacles to the mechanized economies of scale favored by the Anglo-Americans.

Instead of the square-grid settlement pattern adopted by neo-European farmers and ranchers from temperate climates, Chicanos utilized the upland Franco-Iberian (and originally Roman) tradition of the riparian long-lot.[16] The long-lot represents a type of cultural landscape compatible with the biogeographical properties of high-altitude, arid-land environments. The cultural ecological advantage of the long-lot is that it provides every family with access to most of the life zones in the locality. Ideally, every family has access to the piñon-juniper woodlands on the mesa tops and foothills for fuelwood and construction; dry land grass prairies for pasture; riparian bottomlands for access to water, fish, cottonwoods, and wetlands; and irrigated bench land meadows for the planting of row crops, pastures, orchards, and subsistence gardens. The riparian long-lot is not just a boundary-setting tradition. It is an ecosystem with multiple life zones and ecotones (transition zones). Many observers have commented that this agricultural settlement pattern is ecologically sustainable and well adapted to the arid land of the Upper Río Grande watershed.[17]

In Spanish, this agricultural landscape is known as an *extensión*. In some areas of the Upper Río Grande it is called a *vara* strip and in other areas it is known as a *suerte*.[18] In the Lower Río Grande Valley of South Texas the long-lot is called a *porción*. The riparian long-lot is a ribbonlike strip of land that extends many miles through varied topographical and biotic zones. The

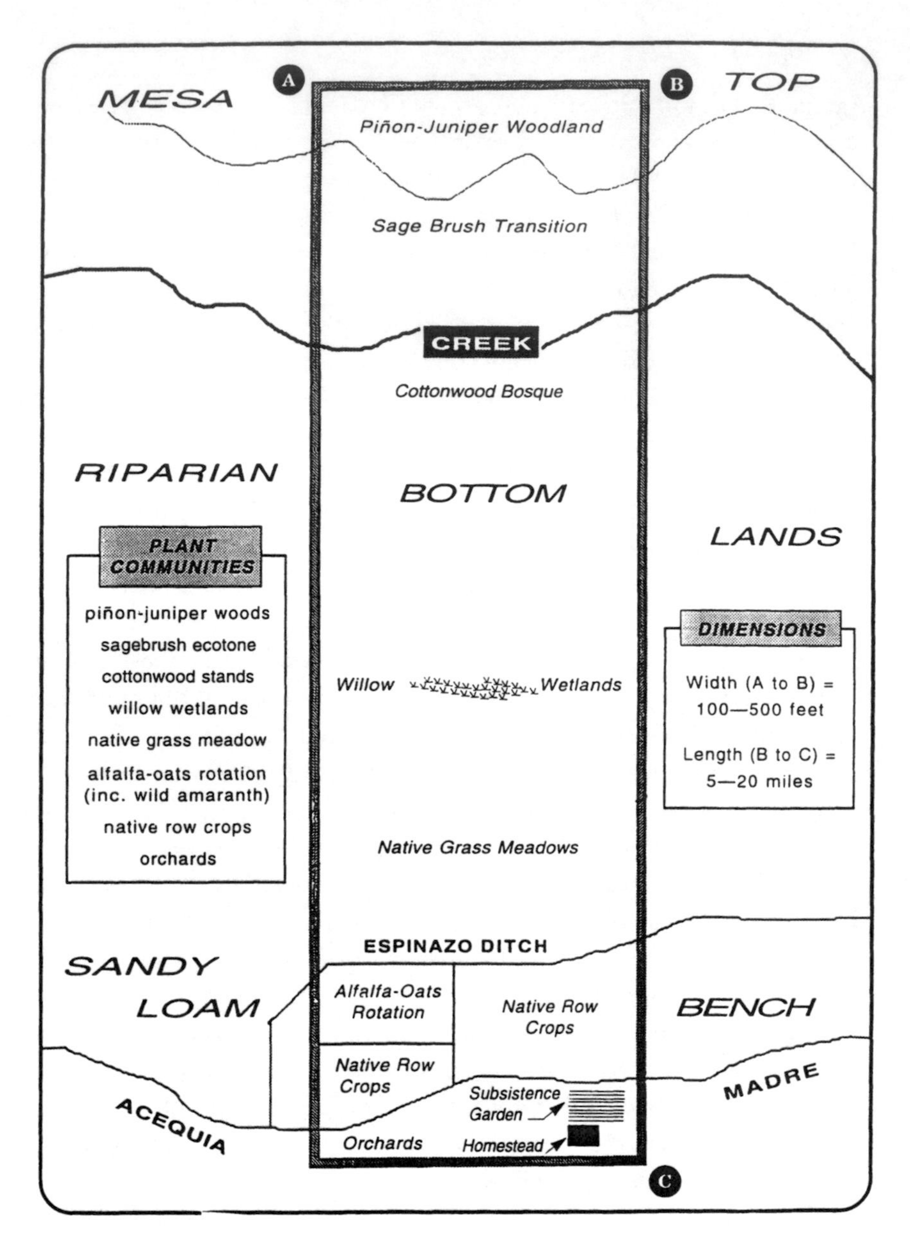

Figure 7.1. The riparian long lot.

size and shape of a long-lot can vary tremendously, depending on microbasin topography, the socioeconomic class standing of the owner(s), patterns of inheritance within families, and the effects of the enclosure of local common and private lands. The width of a long-lot can range from a little less than one hundred to as many as five hundred varas (one vara is equal to 33.3 inches).[19] The length of the long-lot is the significant factor in this cultural landscape. Estimates on the traditional length of historical long-lots vary: five to six miles, as suggested by Wilson and Kammer (1990), fifteen to twenty miles according to Stoller (1993), and ten miles in Carlson (1967).

Acequia Irrigation Systems

As we have seen, acequias (gravity-driven, earthen-ditch irrigation systems) are an integral part of rural Chicano communities, but they are also part of complex agroecosystems. In addition to delivering water to the irrigated fields and pastures, acequias fulfill a variety of ecological functions. What is most striking to us about the ditches is that they fulfill human objectives while simultaneously meeting the needs of wild plants and animals. This is an intrinsic conservation feature of Chicano agroecosystems that is often misrecognized by water engineers and environmentalists as an inefficient and wasteful use of water (see Peña 1995a; see also Gallegos 1998; García 1998; Peña 1998). Because the earthen ditches leak water into the land around them, they are associated with the water-loving phreataphytes (trees and shrubs with extensive root systems like cottonwoods and willows). This means that acequias increase biodiversity by contributing to the creation of wildlife habitats and biological corridors.

From the vantage point of agricultural energy systems, acequias are perhaps the most efficient of all arid-land irrigation technologies (Hall et al. 1979: 29–44). Unlike the mechanical center-pivot sprinkler systems favored by agribusiness monocultures, gravity-ditch systems do not require fossil fuel inputs. Mechanical irrigation systems utilize a great deal of energy yearly, mostly in the form of diesel fuel to power the deep-well pumps that deliver groundwater to the sprinklers. Annual fuel costs for these mechanized systems can run as high as $10,000. In contrast, annual fuel costs for acequias are close to zero.

Another aspect of energy comparison is the nature of trophic and nutrient cycles. The combination of the riparian long-lot with the acequia system contributes to the maintenance of the trophic complexity of the ecosystem by encouraging an optimum mix of relationships among livestock, wildlife, crops, weeds, trees, shrubs, insects, and pathogens. The mechanized system, in contrast, disrupts the trophic webs by homogenizing the landscape and eliminating habitat niches and biological corridors. The acequias actually enhance the flow of energy circuits through the expanded interaction of land, water, flora, and fauna. Mechanized irrigation interrupts these trophic circuits by imposing uniform monocultures on naturally diverse landscapes. Finally, the long-lot/acequia complex also reduces energy inputs by relying on relatively self-enclosed nutrient cycling—that is, all nutrient requirements are met by in situ components of the soil and biota. The mechanized irrigation systems are characterized by open-nutrient cycling: They require high inputs in the form of agroindustrial chemical supplements (fertilizers, herbicides, pesticides, and the like).

Clustering of Wildlife Habitats and Farming Landscapes

The riparian long-lot cultural landscape is characterized by an extraordinary level of biological diversity. The landscape itself, because it includes different life zones, is supportive of an incredible variety of wild plants and animals. In this form, the vara strip agricultural landscapes serve as wildlife habitats and biological corridors linking diverse habitat islands in a given microbasin. One characteristic of Chicano agroecosystems is the existence of amorphous boundaries between natural and cultural landscapes. The boundaries between pastures and wildlife habitat are less definite in this system. In contrast, mechanized monocultures reduce biodiversity because they homogenize and separate the natural and cultural landscapes.

The significance of this landscape clustering feature of Chicano agroecosystems is implied in observations made by a variety of conservation biologists. For example, Reed Noss, a key figure in the field of island biogeography, notes:

> The only success stories in real multiple-use conservation are a handful of indigenous peoples who have somehow been able to coexist with their environments for long periods without impoverishing them. Some indigenous cultures have even contributed to the biodiversity of their regions . . . suggesting that humans have the potential to act as a keystone species in the most positive sense. The beaver provides a good model of how humans could contribute to native biodiversity by creating habitats used by many different species. (Noss 1994:37)

I would like to suggest that Chicano agroecosystems, based as they are on the riparian long-lot/acequia complex, constitute one such example of an indigenous cultural practice that contributes to biodiversity through the protection of the natural landscape integrity of the watershed ecosystem.

Heirloom Crops

Chicano agroecosystems are also characterized by the farmers' preference for native, locally adapted crops. Few Chicano farmers produce hybrid crops. There is an extraordinary range of landraces grown, which are also usually family heirloom crop varieties. The use of landraces means that Chicano farmers are conserving the genetic diversity of food crops and encouraging the adaptation of these varieties to local climatic conditions. This also means that Chicano farmers do not have to utilize high-cost inputs like agroindustrial chemical fertilizers, herbicides, or pesticides. These native plant species are naturally resistant to pathogens and, in some cases, drought. Moreover, because the plants produce fertile seeds, the farmers do not have to rely on seed merchants for their annual seed stocks. The average Chicano farm has a considerable amount of native crop biodiversity (table 7.2). This crop biodiversity eliminates the need for chemical inputs, provides for natural pest and weed controls, and encourages intercropping practices that are beneficial to the soil and its nutrient cycles.

In contrast, the agroindustrial monocultures rely on sterile hybrids for their seed stocks (which makes them dependent on seed merchants and suppliers). Hybrids typically require high inputs to attain higher yields: These inputs include agricultural chemicals (pesticides, herbicides, and fertilizers) and considerable quantities of water for irrigation.[20] The use of hybrids also tends to be associated with the erosion of crop genetic diversity. And there are usually additional impacts involving higher rates of soil erosion, salinization, and compaction.

Some of the landraces that are characteristic of Chicano agroecosystems include *maize de invierno* (a white roasting corn used in making *chicos* or *posole*), *bolitas* (a beige-colored bean related to the pinto), a wide variety of *chiles* (hot green peppers), and *calabasita* (Mexican green squash). Various "naturalized" exotic varieties have been adapted to the high-altitude conditions of the Culebra watershed; *habas* (horse beans) are one such example. Historically, Chicano agroecosystems tend not to have extensive plots of land dedicated to alfalfa or other field crops for feeding livestock. Pastures with perennial polycultures of native grass are preferred. The Chicano farmer favors a combination of landraces for row crops, native grasses for pasture, and a variety of imported hybrids and exotic landraces for subsistence gardens. In

Table 7.2. Domesticated crops, animals, and grasses in Hispano agroecosystems.

Crops		
beans (*bolita*)*	chives	parsley
beans (pinto)*	cilantro	peas (English)
beans (string)	corn (blue)	peas (sweet)
broccoli (3)	corn (yellow sweet)	potatoes (5)
cabbage (2)	decorative flowers (gladiolus, etc.)	turnips
*calabasita**		vine tomatoes (5)
carrots (3)	*habas* (horse beans)*	wheat (3)
cauliflower	lettuce (3)	yellow crookneck squash
chicos (7)*	onions	zucchini
Orchard Crops		
apple (2)	chokecherry*	plum (3)/(1*)
cherry (2)	pear	
Animals		
cats	ducks (5)	hogs (6)
cattle (6)	geese (3)	horses
chickens (8)	goats (5)	rabbits
dogs	guinea fowl	sheep (4)
Grasses		
black grama*	brohme*	wheat (luna crested)
blue fescue*	redtop*	wheat (western crested)
blue grama*	timothy*	

Notes: Numbers in parentheses indicate horticultural varieties and breeds. Asterisks indicate native landraces.

addition, Chicano agroecosystems typically include orchards with exotic fruit trees (apple, pear, plum, and cherry), imported berry brambles (raspberry, currant), native berry shrubs (chokecherry), and a wild (semidomesticated) miniature plum known locally as *cirhuelita del indio* (see table 7.2 for a partial list of crops grown on Chicano acequia-fed farms).

Natural Pest and Weed Controls

Given the biodiversity of crops grown in Chicano agroecosystems, it is not surprising that natural (biological) pest and weed controls are the order of the day. The primary form of weed and pest control involves careful intercropping of landrace crops. Intercropping, combined with rotational plantings, creates a condition known as allelopathy (what the home gardener knows as "companion planting" and ecologists recognize as a chemical interrelationship between plants). For example, the traditional trinity of Indian crops—corn, beans, and squash—serves more than to provide for a balanced diet. Together, these companion plants work to fertilize the soil (beans as legumes are nitrogen-fixers), control weeds and soil erosion (squash as a ground cover reduces weed invasions and soil loss), and eliminate many insect pests (the biodiversity and adaptation of the three crops to local microclimate and soil conditions help them resist diseases and infestations). (For further discussion, see Altieri et al. 1987; see also Barreriro 1992).

Chicano agroecosystems can thus be characterized as landrace polycultures that feature both species and structural diversity. They exploit the full range of microenvironments, maintain and enhance nutrient cycles and soil tilth, rely on biological interdependencies that provide pest control, rely on local resources with little mechanical technology, and rely on local varieties of crops and incorporate wild plants and animals (see also Altieri et al. 1987, Harlan 1976).

Holistic Land and Livestock Management

One of the most significant, and most often overlooked, characteristics of Chicano agroecosystems is their integration of farming and ranching. As noted earlier, Chicano agroecosystems are not just farms and not just ranches: They typically incorporate aspects of both production systems—hence the term "agropastoral." The typical Chicano agropastoral operation produces at least four types of plant crops: row crops (such as corn, beans, squash, chili), forage crops (such as alfalfa, hay), pastures (native grasses such as timothy, blue fescue, redtop, brome), and subsistence garden crops (such as corn, beans, squash, chili, tomatoes, peas, broccoli). But these operations also produce livestock, typically, cattle, sheep, goats, pigs, and horses. This integration of crops, forage, pasture, and livestock increases biodiversity and maintains trophic complexity. The presence of farm animals also means that a steady supply of organic fertilizer is available in the form of manure. Sheep and goats can be used to control invasive noxious weeds.

Land and range management practices in Chicano agroecosystems are centered on controlling three types of problems: overgrazing, loss of soil fertility, and soil erosion. The use of Holistic Resource Management (HRM) practices to control grazing and soil erosion is increasingly evident (on HRM, see Savory 1988). The HRM model involves several primary practices: rotational grazing to reduce pressures on forage and pasture crops; electrical paddocks to control livestock movements and concentrations; and intense supervision of grazing animals.[21]

Polycultures and Rotational Intercropping

Chicano agroecosystems combine elements of both perennial and annual polycultures. The perennial polycultures include native grass meadows that are never tilled or cultivated. These meadows are used as rotational pastures for grazing livestock. The annual polycultures include row crops that can be intercropped (with the corn-bean-squash-chili complex being the most common). The row crop plantings usually involve minimum tillage and plowing.[22] Some Chicano agropastoralists have in more recent times adopted monoculture plantings of alfalfa and other forage and livestock crops. However, Chicanos usually avoid alfalfa monocultures and, in most cases, plantings follow eight- to twelve-year rotational sequences: for example, alfalfa-oats-barley-corn. Rotations often include a fallow period.

Soil and Water Conservation

Chicano agroecosystems have historically experienced fewer problems with soil erosion than have agroindustrial monocultures. There are several features that contribute to soil conservation. First, most Chicano agropastoralists practice zero or minimum tillage, particularly in the native grass meadows that predominate in the riparian bottomlands. Second, where tillage and cultivation are practiced, the combination of crop diversity and cover crops reduces soil erosion. Third, since most agropastoralists avoid large-scale mechanization, there are fewer erosive impacts from the use of heavy machinery. Fourth, most agropastoralists practice "organic" farming, with very little use of agroindustrial chemicals; organic farming practices tend to increase soil tilth and reduce soil erosion.

There are some other factors that contribute to soil conservation on Chicano farms and ranches. Historically, since Chicanos were the first to settle in their respective microbasins, they tend to have the best land; few Chicanos have been pushed off onto marginal lands that are more erosive. The tendency of Chicanos has been to farm only in riparian bottomlands: These areas have deeper soil horizons (usually dating to Pleistocene deposition) and tend to be protected from wind erosion by the proximity of higher lands (such as surrounding mesas and foothills). Another contributing factor to soil conservation is the existence of numerous windbreaks. The acequia networks, as we have seen, create numerous cottonwood and willow stands that double as windbreaks. Tree lines, woodlots, orchards, and naturally occurring wetland willow and cottonwood stands provide further protection against wind erosion.

Under certain circumstances, acequia irrigation practices can contribute to soil erosion. Such circumstances nearly always involve human error: For example, flooding fields with excessive water or irrigating at too rapid a pace can contribute to soil erosion from runoff. In some cases, farmers may inappropriately try to irrigate fields with too steep a gradient. If fields are furrowed at angles parallel to the gradient, erosion may result. However, the main cause of soil erosion in Chicano agroecosystems has been overgrazing.[23] Overgrazing became a problem only after the conquest of the bioregion by the United States: The commercialization of livestock production, the expanded demand for beef occasioned by the arrival of the railroad, and the opening of new markets were major factors in the overgrazing of these lands (see Peña 1992, 1994). More recently, Chicanos have adopted a variety of strategies to control grazing (for example, HRM as noted above).

The establishment of the Soil Conservation Service (SCS, now the Natural Resources and Conservation Service NRCS), has, on the whole, proven beneficial to Chicano practices. However, the SCS has a mixed record in attending to the needs of Chicano agropastoralists. Like the Extension Service, the SCS has not always placed a high priority on the needs of Chicano agricultural regions, and there are some cases where the SCS encouraged Chicanos to destroy woodlands in order to expand the acreage under cultivation with alfalfa monocrops. In more recent years, the SCS has increased the number of Chicano staff and emphasized projects in "limited resource" farming communities. Some of the more interesting projects include the introduction of regenerative and restorationist projects to assist locals in repairing damaged lands.

Water conservation is another aspect of Chicano agroecosystems that merits discussion. The acequia irrigation tradition has been criticized as wasteful and inefficient. We have seen how this criticism is most often made by state hydrologists and some environmentalists. The debate has raged for decades, with critics emphasizing the "loss" of water, since most acequias are earthen works. This has led to pressure to line the ditches with concrete to prevent the leakage of water. But the "loss" of water is a matter of perspective: The water is lost to what legal experts call "beneficial human use." From an ecological perspective, the leaking water is not lost. We have already noted that acequias create habitat niches and biological corridors and thus contribute to the maintenance of biodiversity. The water "lost" by leaking acequias is very much a part of the local hydrological cycle. The water returns to the cycle via evapotranspiration (the evaporation of water through plant life) and aquifer recharge. These processes can contribute to cooler and wetter local microclimates: Evapotranspiration, for example, contributes to local rain cycles through convection currents that result in summer afternoon thunderstorms.

Low Mechanization/Human and Animal Power

Chicano agroecosystems are characterized by low levels of mechanization. The cultural landscape of the riparian long-lot does not lend itself easily to extensive mechanization: Most long-lots are too narrow for the use of large machinery like combines or center-pivot sprinkler systems. The huge capital expenditures required for large machinery further discourage mechanization on most Chicano farms and ranches. Historically, Chicano farmers and ranchers have relied on human and animal power for their plowing, planting, cultivating, and harvesting. As long as family labor is available for farmwork, the incentive to mechanize remains low. However, increasing mechanization is apparent on some farms and ranches. The use of tractors, moldboard plows, windrowers (swathers), and bailers is not altogether uncommon, particularly among Chicanos who produce alfalfa and hay for livestock feed. But, to the extent that farmers adopt perennial polycultures, machinery is less likely to be used as extensively.

Restoration Ecology

Given historical problems with overgrazing and soil erosion, some Chicano farmers and ranchers have adopted regenerative and restorative agricultural practices to repair damaged lands. One type of practice involves the restoration (really rehabilitation) of the native dry land grass prairie ecology that was predominant in much of the bioregion before the advent of the railroad and the commercial raising of livestock. The native blue grama prairies were overgrazed in much of the bioregion between the 1890s and 1930s (see Peña 1995a; Peña forthcoming; Peña and Martínez 1998). Restoration work in the Culebra microbasin is becoming more common as Chicano farmers, working with the SCS and other agencies, re-establish prairies using a combination of native and exotic dry land grasses.

Autarkic Prosumption

Perhaps one of the reasons for the endurance of Chicano farms and ranches is that they have always produced both for subsistence and the market. Production for subsistence, or prosumption, has stabilized the farming operations during bust cycles in the economy, insulating the smallholders against the loss of land and keeping them in agricultural production. During periods of rising market demand, Chicanos have responded by producing and delivering farm produce to the market (in this region, particularly the organic produce markets in northern New Mexico). This autarkic quality has allowed Chicano agropastoralists to survive the boom/bust cycles of the economy. Moreover, when market conditions have been poor, the local producers have turned to traditional bartering networks.

Common Property Resources

Chicano agroecosystems have traditionally relied on access to the common lands of community land grants. Access to common property resources is a critical component in the sustainability of the agropastoral tradition. The availability of common lands for limited rotational grazing, wood gathering for fuel, hunting, fishing, and wildcrafting (the harvesting of edible wild plants and medicinals) has helped to stabilize Chicano agroecosystems and reduce land degradation on the private riparian long-lots. However, enclosure of these common lands has proven profoundly detrimental to the agropastoralists. In the Upper Río Grande, enclosure has been practically universal and it has destroyed the ability of many families to remain in agriculture. The restoration of common lands is thus one of the most important unresolved issues facing the ethnoecology of Chicano farming communities. Restoration, in this context, is twofold: It involves both the ecological restoration of degraded common lands and the restoration of traditional usufructuary rights (for further discussion, see Peña 1995a; Peña and Valdéz Mondragon 1997).

Links to Market and Governmental Institutions

Chicano agropastoralists have always produced for the market and not just for subsistence; this is why there was a long tradition in the bioregion involving the construction of *carretas* (carts), which were used to transport farm produce to the market. There was a long period, from 1848 through the 1970s, when Chicano agropastoralists experienced discrimination in credit markets. Many Chicanos were denied credit by banks and other agricultural production creditors. However, this may have proven to be a blessing in disguise. Since Chicanos could not gain access to credit, they avoided debt and thus the loss of land that is often associated with indebtedness. Since the late 1970s, Chicano agropastoralists have enjoyed relatively unfettered access to credit markets and have demonstrated their ability to use it to their advantage. It is now not uncommon for Chicanos to make use of producer credit associations and federal and private banks to expand their operations or acquire new land (see Peña forthcoming).

Like non-Chicano producers, Chicanos are establishing relationships with a variety of governmental agencies to improve and strengthen their agricultural operations. In addition to relationships with private and public sector creditors, Chicanos are working with a full range of governmental agencies such as the Extension Service, NRCS, Agricultural Stabilization and Conservation Service (ASCS), and U.S. Forest Service (USFS). Chicanos have played a key role in the establishment of soil conservation districts and are active in federal projects like the Conservation Reserve Program (CRP), designed to protect wetlands and other landscapes that provide wildlife habitat.

Conclusion

The ethnoecology of Chicano farming systems in the Culebra microbasin is characterized by the riparian long-lot/acequia cultural landscape. This ethnoecological complex promotes and protects biodiversity and represents a sustainable adaptation to local environmental and cultural conditions in high-altitude arid-zone watersheds. As an autochthonous form of local democratic self-governance, the watershed commonwealth serves two primary roles in the agropastoral community: "technical" (as in the maintenance and operation of the ditch networks) and "ethical" (as in the transgenerational reproduction of land and water conservation values). Finally, agroecological practices derived from local knowledge utilize renewable energy systems, mimic natural patterns in their species and structural diversity, preserve the diversity of heirloom germplasms, contribute to a local sense of place and land ethics, and contribute to sustainable patterns for agriculture within the regional, political, and economic context.

Cultural landscapes in Chicano agroecosystems—that is, the riparian long-lot\acequia complex—clearly present a unique set of opportunities for the protection of biological diversity. For example, from the perspective of island biogeography we might argue that these agroecosystems serve as habitat islands and biological corridors connecting larger regional islands. Under these circumstances, farming and ranching are directly productive of biodiversity because the land use pattern encourages the protection of an optimum mix of plant and animal communities. The practices that sustain the land and water also provide stability for the agropastoral community.

There are many serious threats to the integrity of these biological island habitats. For example, the Pecos and Wheeler Peak Wilderness areas have been severely damaged by overgrazing and excessive recreational use. The Taos ski area presents a threat to the Wheeler Peak microbasin, while a molybdenum mine in Questa threatens the Red River and Latir Peaks areas. Battle Mountain Gold presents mining threats in the Rito Seco area of the Culebra microbasin, and the Forbes Trinchera, with its three subdivisions, presents a threat to the watershed from real estate development and the four hundred miles of roadway associated with widespread construction and timber operations. The Culebra Mountain Tract is currently threatened by massive logging activities involving cuts of eighty to one hundred million board feet on thirty-four thousand acres (see Peña and Valdéz Mondragon 1997). Nevertheless, the Culebra remains the primary undisturbed area between the northern and southern Sangre de Cristo. It is the only remaining, relatively intact, habitat island in the mountain corridor without protection against development and environmental degradation.

The struggle to protect the Culebra as a biological corridor between the southern and northern Sangre de Cristo mountains is developing in the context of a campaign for the preservation of rare and endangered cultural landscapes. The most vulnerable aspect of this ethnoecological complex is the need for a healthy, undisturbed watershed. Therefore the most critical public policy and organizing challenges for the farming communities of the Culebra microbasin center on land reform (that is, the restoration of a common property regime) and environmental degradation (a result of the enclosure of the commons).[24] At stake in this struggle is the preservation of a national environmental treasure and the survival of a human community that has evolved into a rare example of a human "keystone species."

Notes

1. This cultural "headwaters" bioregion includes the counties of Río Arriba, Taos, Mora, San Miguel, and Guadalupe in New Mexico, and Costilla and Conejos in Colorado. This roughly covers the distance from San Luis, Colorado, to Española, New Mexico, (approximately 130 river miles).

2. More than 99 percent of the water supply in the Upper Río Grande comes

from this headwaters bioregion in southern Colorado and northern New Mexico. Most of the water is runoff from melting snow in the high mountains. Technically, most of Mora and San Miguel Counties are in the Arkansas and Pecos River watersheds, but these are also predominantly Chicano areas and share cultural and familial ties with the rest of the "Rio Arriba." See Hay (1963:491).

3. The oldest Spanish-Mexican settlement in the Upper Río Grande was San Gabriel (settled in 1598). The oldest existing settlement is Santa Fe (settled in 1610). Most of the settlers who came to work on the land were *mestizos* (the offspring of Indian and Spanish mixtures). The Spanish-speaking peoples of the Upper Río Grande are thus primarily the descendants of indigenous mestizos and not full-blooded Spaniards. We prefer to use the term Chicano (instead of Spanish-American or Hispanic) in order to acknowledge the diverse character of the mestizo culture, which has roots in Mexican, native American, Iberian, and Moorish (north African) cultural traditions.

4. The agropastoral upland village is based on the integration of farming and livestock-raising. For more on the cultural ecology of Chicano agropastoralism see Peña (1992); Peña (forthcoming); Van Ness (1987).

5. See García (1989, 1998); Peña (1992, 1998); see also the intriguing commentary by Noss (1994).

6. Technically, the Culebra microbasin is not considered tributary to the Río Grande. Since the turn of the century (1910), when Euroamerican farmers constructed reservoirs in the watershed, the Río Culebra has not normally reached the Río Grande, twenty miles west, as a surface flow. However, the watershed is still connected to the Río Grande via groundwater aquifers.

7. The villages of the Culebra include Viejo San Acacio (settled temporarily in 1850 and permanently in 1853), San Luis (1851), San Pablo (1852), San Pedro (1852), San Francisco (1854), Chama (1855), and Los Fuertes (1860?). On the settlement of the Sangre de Cristo land grant, see Stoller (1992) and Valdéz Valdéz (1991).

8. In addition to the twenty-three acequias with original nineteenth-century water rights, there are another forty-five acequias in the Culebra microbasin with more junior surface water rights.

9. In addition, the BMG strip mine and cyanide leach vat processing mill is located on land abutting the Mountain Tract in the Rito Seco watershed; see Peña and Gallegos (1993). Since 1995, logging on the Taylor Ranch threatens the watershed with cuts of 90 to 210 million board feet.

10. Flooding is especially a problem during the "rain-on-snow" events characteristic of the area during the wet months of spring through early summer. Rain accelerates the rate at which snow pack melts, especially in exposed cut block areas.

11. For a scientific study of the impact of logging operations on the Culebra watershed see Curry (1995); see also Peña and Valdéz Mondragon (1997).

12. For more on the land rights struggle in San Luis, see La Sierra Foundation of San Luis (1995); Peña (1995a, 1998); Peña and Gallegos (1993, 1997); Peña and Valdéz Mondragon (1997); Stoller (1985).

13. For an overview of the principles of agroecology see Altieri et al. (1987).

14. Much of previous research has been done at the level of regions, communities, or land grants and not specific farms and ranches. Moreover, previous research is based on cultural-ecological and not ethnoecological principles.

15. This research is funded by a four-year grant from the National Endowment for the Humanities (NEH) with matching support from the Colorado College. The NEH study, "Upper Río Grande Hispano Farms: A Cultural and Natural History of Land Ethics in Transition, 1598–1998," is coordinated by the Río Grande Bioregions Project, a research unit of the Hulbert Center for Southwestern Studies at Colorado College. The first publication associated with this project is Peña (1998).

16. The Metis of Canada fought a war (in 1868–1870 and 1885) with the British over the imposition of the square-grid land survey system. The Metis, of French/Native Canadian mixture, were one of the few other major ethnic communities, besides the Chicanos, to make use of the riparian long-lot in North America. The Metis ultimately lost this struggle, while the Chicano cultural landscapes endured. See Howard (1952); Powell (1962:34,n.7).

17. The first to make this argument was John W. Powell (1890:111–16); see also Peña (1992); Rívera and Peña (1997); Van Ness (1987).

18. In the San Luis Valley, the long-lot is called a *vara* strip or *extensión;* in the Embudo-Velarde-Alcalde region it is called a *suerte.* The author thanks Estevan Arellano for this clarification on the regional nuances of the local vernacular terms used to describe this cultural landscape.

19. The vara is a unit of measurement used in the ancient metes and bounds system; one vara is approximately 33.3 inches wide. The vara measures width and not length. In the Iberoamerican system, length is measured by leagues (*leguas*). See Van Ness and Van Ness (1980:9).

20. We studied records in the State Engineers Office in Denver and found that center-pivot sprinkler systems use three to five times as much water as traditional acequia systems in the San Luis Valley.

21. It is interesting to note that this aspect of ranching activity has benefited from the simple technological addition of the truck. With a truck, one person can easily supervise a herd of two hundred to five hundred animals.

22. For example, during a soil survey conducted in July 1994, Robert Curry (staff watershed scientist) found that the soil horizon at the Corpus A. Gallegos Ranches in San Luis yielded an "A" horizon at least five feet deep. The only evidence of a "plow pan" was a half-inch-thick clay lens at about two feet in one of the corn *milpas.* See also Peña (1995b).

23. See Peña (1994) for more on the problem of overgrazing in Chicano farming communities.

24. The problem of land degradation is complicated by the complete enclosure of the common lands, an issue we explore elsewhere. See Peña (1995a); Peña and Martínez (1998).

References

Altieri, M., S. Hecht, and R. Norgaard. 1987. *Agroecology: The Scientific Basis of Alternative Agriculture.* Boulder: Westview Press.

Barreriro, J., ed. 1992. "Indian Corn of the Americas: Gift to the World." *Northeast Indian Quarterly* 6(1/2):4–96.

Carlson, A. W. 1967. "Rural Settlement Patterns in the San Luis Valley." *Colorado Magazine* 44(2):111–28.

Costilla County Conservancy District. 1993. "Revenue Potential and Ethical Issues in the Management of the Culebra Mountain Tract as a Common Property Resource. San Luis: Costilla County Conservancy District." Prepared by Devon G. Peña and Robert K. Green.

Curry, R. 1995. "State of the Culebra Watershed. I: The Southern Tributaries." *La Sierra: National Edition,* 1(fall 1995/winter 1996): 9–11.

Gallegos, J. 1998. "*Acequia* Tales: Stories from a Chicano Centennial Farm." In *Chicano Culture, Ecology, Politics: Subversive Kin.* D. G. Peña, ed. Tucson: University of Arizona Press.

García, R. 1989. "A Philosopher in Aztlan: Notes Toward an Ethnometaphysics in the IndoHispano (Chicano) Southwest." Unpublished doctoral diss. Boulder: Department of Philosophy, University of Colorado.

———. 1998. "Notes on (Home)Land Ethics: Ideas, Values, and the Land." In *Chicano Culture, Ecology, Politics: Subversive Kin.* D. G. Peña, ed. Tucson: University of Arizona Press.

Hall, A., G. H. Cannell, and H. W. Lawton, eds. 1979. *Agriculture in Semi-Arid Environments: Ecological Studies 34.* Berlin: Springer-Verlag.

Harlan, J. R. 1976. "Genetic Resources in Wild Relatives of Crops." *Crop Science* 16(3):329–33.

Hay, J. 1963. "Upper Río Grande: Embattled River." In *Aridity and Man: The Challenge of the Arid Lands in the United States.* C. Hodge, ed. Washington, D.C.: American Association for the Advancement of Science.

Howard, J. K. 1952. *Strange Empire.* New York: Morrow.

Jones, J. A., and G. E. Grant. 1996. "Peak Flow Responses to Clear-cutting and Roads in Small and Large Basins, Western Cascades, Oregon." *Water Resources Research* 32(4):959–74.

La Sierra Foundation of San Luis. 1995. *A Proposal for the Establishment of a Community Land Trust Based on the Principles of Environmental Justice.* San Luis: La Sierra Foundation of San Luis.

Noss, R. 1994. "A Sustainable Forest is a Diverse and Natural Forest." In *Clearcut: The Tragedy of Industrial Forestry.* W. DeVall, ed. San Francisco: Sierra Club.

Pearson, M. 1994. "La Sierra Tract: A Key Link in the Landscape of the Southern Rockies." *La Sierra* 1(1):7–8.

Peña, D. G. 1992. "The 'Brown' and the 'Green': Chicanos and Environmental Politics in the Upper Río Grande." *Capitalism, Nature, Socialism* 3(1):79–103.

———. 1993. "Acequias and Chicana/o Land Ethics." Paper presented at the 35th annual Conference of the Western Social Science Association. Corpus Christi, Tex. (April).

———. 1994. "Pasture Poachers, Water Hogs and Ridge Runners: Archetypes in the Site Ethnography of Local Environmental Conflicts." Paper presented at the 36th annual Conference of the Western Social Science Association. Albuquerque, N.Mex. (April).

———. 1995a. *Progress Report: Research Findings from the Upper Río Grande Hispano Farms Study. Vol. I: Summary of Environmental History Research Modules (June 1994–May 1995)*. Colorado Springs: Río Grande Bioregions Project.

———. 1995b. "Environmental History of the Culebra Watershed." Field report, Upper Rio Grande Hispano Farms Study, Rio Grande Bioregions Project. Colorado Springs: Colorado College.

———, ed. 1998. *Chicano Culture, Ecology, Politics: Subversive Kin*. Tucson: University of Arizona Press.

———. (forthcoming). *Gaia in Aztlán: The Politics of Place in the Río Arriba*. Unpublished manuscript. Colorado Springs: Department of Sociology, Colorado College.

Peña, D. G., and J. Gallegos. 1993. "Nature and Chicanos in Southern Colorado." In *Confronting Environmental Racism: Voices from the Grassroots*. R. Bullard, ed. Boston: South End Press.

———. 1997. "Local Knowledge and Collaborative Environmental Action Research." In *Building Community: Social Science in Action*. P. Nyden et al., eds. Thousand Oaks, Calif.: Pine Forge Press.

Peña, D. G., and R. O. Martínez. 1998. "The Capitalist Tool, the Lawless, and the Violent: A Critique of Recent Southwestern Environmental History." In *Chicano Culture, Ecology, Politics: Subversive Kin*. D. G. Peña, ed. Tucson: University of Arizona Press.

Peña, D. G., R. O. Martínez, and L. McFarland. 1993. "Rural Chicana/o Communities and the Environment: An Attitudinal Survey of Residents of Costilla County, Colorado." *Perspectives in Mexican American Studies* 4(August):45–74.

Peña, D. G., and M. Valdéz Mondragon. 1997. "The 'Brown' and the 'Green' Revisited: Chicanos and Environmental Politics in the Upper Rio Grande." In *The Quest for Ecological Democracy: Movements for Environmental Justice in the United States*. D. Faber, ed. New York: Guilford Press.

Powell, J. W. 1890. "Institutions for the Aridlands." *Century Magazine* 40(May–October):111–16.

———. 1962. *Report on the Lands of the Arid Region of the United States*. Cambridge: Belknap Press of Harvard University Press.

Reynolds, R. T. 1990. "Distribution and Habitat of Mexican Spotted Owls in Colorado: Preliminary Results." Unpublished report. Laramie, Wyo.: Rocky Mountain Forest and Range Experiment Station.

Rívera, J. A. 1997. *Water and Democracy in the Southwest: The Acequia Papers*. Unpublished manuscript.

Rívera, J. A., and D. G. Peña. 1997. "Historic Acequia Communities in the Upper

Río Grande: Policy for Cultural and Ecological Protection in Arid Land Environments." Unpublished manuscript presented to the Rural Latino Studies Working Group.
Savory, A. 1988. *Holistic Resource Management*. Washington, D.C.: Island Press.
Stoller, M. 1985. "La tierra y la merced." In *La cultura contante de San Luis*. R. Tweeuwen, ed. San Luis: San Luis Museum and Cultural Center.
———. 1992. *"The Settlement History of the San Luis Valley."* Unpublished manuscript. Colorado Springs: Department of Anthropology, Colorado College.
———. 1993. *Preliminary Manuscript on the History of the Sangre de Cristo Land Grant and the Claims of the People of the Culebra River Villages on Their Lands*. Colorado Springs: Department of Anthropology, Colorado College.
Valdéz, A., and M. Valdéz. 1991. *The Culebra River Villages of Costilla County: Village Architecture and Its Historical Context, 1851–1940*. Denver: Colorado Historical Society.
Van Ness, J. R. 1987. "Hispanic Land Grants: Ecology and Subsistence in the Uplands of Northern New Mexico and Southern Colorado." In *Land, Water, and Culture: New Perspectives on Hispanic Land Grants*. C. L. Briggs and J. R. Van Ness, eds. Albuquerque: University of New Mexico Press.
Van Ness, J. R., and C. M. Van Ness, eds. 1980. *Spanish and Mexican Land Grants in New Mexico and Colorado*. Manhattan, Kans.: Sunflower University Press.
Webb, B. 1983. "Distribution and Nesting Requirements of Montane Forest Owls in Colorado." Part IV: "Spotted Owl *(Strix occidentalis)*." *Colorado Field Ornithologists Journal* 17(1):2–8.
Wilkinson, C. H. 1988. *The Eagle Bird: Mapping the New West*. Tucson: University of Arizona Press.
———. 1992. *Crossing the Next Meridian: Land, Water, and the Future of the West*. Washington, D.C.: Island Press.
Wilson, C., and D. Kammer. 1990. *Community and Continuity: The History, Architecture and Cultural Landscape of La Tierra Amarilla*. Santa Fe: New Mexico Historic Preservation Division.
Worster, D. 1994. *An Unsettled Country: Changing Landscapes of the American West*. Albuquerque: University of New Mexico Press.

8

Conserving Folk Crop Varieties

Different Agricultures, Different Goals

DANIELA SOLERI
STEVEN E. SMITH

Over the last twenty-five years, a substantial investment in the conservation of crop genetic resources worldwide has been made by public and private sectors in many countries. One impetus for this conservation effort came from researchers who were alarmed at the rapid disappearance of folk crop varieties (also known as landraces) as well as the habitat of wild crop relatives (see, for example, Frankel 1970). Another motivation for crop genetic resource conservation came from observing the effect of a narrowed genetic base in the agricultural systems of industrialized countries. An example of this was the southern corn leaf blight damage in the United States in 1971, a loss of an estimated $500 million to $1 billion, or about 15 percent of the U.S. crop that year, attributed to the broad use of the same cytoplasmic male sterility gene (Walsh 1981). The result was increased concern for genetic resources and diversity that continues today under the broader title of biodiversity.

A new issue is emerging in the discussion of crop genetic resource conservation, or perhaps more correctly, it is just now beginning to be articulated (Soleri and Smith 1995), concerning the goals of conservation and some of their implications. This paper discusses the conservation of crop genetic resources, specifically folk varieties, in terms of the genetic goals of the people for whom those resources are being conserved, how different goals require different conservation strategies, and some ideas on how we can start to conduct research that specifically addresses those goals. We look particularly at farmer-breeder selection and the characteristics of folk crop varieties for insights into the genetic goals for traditional crop varieties and the people who cultivate them.

Goals of Folk Variety Conservation

The genetic diversity of crop genetic resources is a valuable part of all agricultural systems. How this diversity is maintained and used differs, however, among different types of agricultural systems. Crop genetic resources, including folk varieties, may be conserved in situ (on site and in the farming system where they were most recently developed) or ex situ (away from that environment, usually in seed banks or occasionally in living collections). Whether in or ex situ, the goals of folk variety conservation differ. While there are certainly diverse goals for the conservation of folk varieties, they all lie on a continuum between the polar perspectives we label "conventional" and "farmer-breeder." Although extreme, these nevertheless represent actual conservation goals. To understand these perspectives and the genetic goals they imply requires placing them within the agricultural systems of which they are a part.

The Conventional Perspective

The "conventional perspective" has been followed by most plant breeders and conservationists concerned with crop genetic resources over the past twenty-five years. From this perspective, crop genetic resources conservation is undertaken to meet the needs of industrial agriculture, typically large-scale, high-external-input agriculture (Plucknett et al. 1987). Industrial agriculture has been based on the creation of improved growing environments and on uniform, energy-intensive cultural practices (soil preparation, irrigation, insect and pathogen [hereafter "pest"] control) extending over large areas. Plant breeding for these agricultural systems has focused primarily on maximizing yields. This objective has been achieved by increasing the harvest index in some species (such as rice), and by exploiting the genotype by environment interaction (G x E) that, when positive, makes some populations highly responsive to the "improved" growing environments that are created with high levels of inputs (Simmonds 1991b). The assumption that these inputs would always be available has meant that local adaptation, particularly to abiotic conditions such as water stress or lack of soil nutrients, has not been an important consideration in most of these breeding programs. Such adaptation may therefore have been lost during breeding in favor of yield.

On the other hand, there has been an interest in specific, usually qualitative traits such as resistance to pests or pathogens. That is, both genetic and chemical controls have been used in response to these problems. However, pests and pathogens may rapidly evolve resistance to either the chemical or genetic controls typically used in industrial agriculture for two reasons. First, the extensive scale of these systems (large fields and often entire regions planted to the same variety or varieties and maintained with closely similar cultural practices), exerts strong selection pressure for the development of resistance (Gould 1988). Second, the vertical (monomorphic, oligogenic) resistance used in many conventional breeding programs also creates strong selection pressure for the development of resistance, and opportunity for the success of that resistance (Simmonds 1991a).

These are some of the reasons that crop varieties produced by conventional breeding efforts (hereafter "modern varieties," including Green Revolution varieties) often require frequent replacement with newly developed modern varieties. This is sometimes referred to as the varietal "relay race" of industrial agriculture (Plucknett et al. 1987:19ff). These varieties are characteristically more homogeneous, and in the case of cross-pollinating species, they may be more homozygous relative to folk varieties of the same species (Frankel and Soulé 1981).

According to the conventional perspective, crop genetic resources are recognized as one of the raw materials essential to the continuing development and replacement of modern varieties. In these agricultural systems at least some genetic diversity is maintained apart from the currently cultivated populations in collections established for conservation. Collections are screened for specific traits and, when possible, the alleles for those traits are then introduced into the elite breeding lines developed by Western-trained, formal plant breeders. This approach is reflected in the responses to a 1986 survey of seventeen U.S. wheat breeders (Cox 1991). Those breeders estimated that approximately 24 percent of the germplasm in advanced breeding lines of *Triticum aestivum* was estimated to come from "exotic" sources (folk varieties and crop relatives). Half of those surveyed intended to increase their use of exotic germplasm, and none planned on decreasing such use; they concluded that within *T. aestivum* breeding programs, folk varieties will be used only for specific problems such as apparent lack of disease or pest resistance.

The genetic goals of folk variety conservation from this perspective concern the genetic content of folk variety populations—that is, the extent of polymorphisms and allelic diversity of those populations. Sampling methods (Marshall and Brown 1975) have been based on maximizing allelic diversity, and storage and regeneration have sought to preserve this diversity. From this perspective, germplasm collections as sources of oligogenic, qualitative traits can never be large enough and should be constantly supplemented to include the products of ongoing coevolution in nature.

The Farmer-Breeder Perspective

Another perspective on folk variety conservation, internal in its orientation, is that the role of these genetic resources within their original, local farming system is an equally valid reason for their conservation (Cleveland et al. 1994). In this paper it is referred to as the farmer-breeder perspective because it attempts to look at the genetic resources in folk varieties from the standpoint of the farmer-breeders (hereafter "farmers") of traditionally based, low-external-input agricultural communities. However, the discussion that follows is a genetic interpretation of our understanding of how these agricultural systems work and does not presume to speak for those farmers. Working with them to support their efforts to identify and communicate their own crop genetic goals is a research priority.

Interest among researchers in the farmers' perspective is based on the purported relationship between traditionally based, low-external-input agricultural systems and particular aspects of natural resource management, agricultural production, and contributions to sociocultural integrity. In plant breeding there is increasing recognition that selection in the local growing environment is the most effective approach to the development of crop varieties that are adapted to those conditions and therefore require a minimum of external inputs to grow and produce a harvest (Ceccarelli et al. 1992a; Simmonds 1991b). Locally developed crop varieties that are genetically diverse (relative to modern varieties), such as many folk varieties, often have greater yield stability (measured as variation in harvest produced from year to year) under low-input conditions than do modern varieties (Cleveland n.d.; Finlay and Wilkinson 1963). To some extent, genetic variation appears to contribute to this stability, for example in the broad or horizontal resistance to local pests or pathogens that may be present in folk variety populations (Simmonds 1991a). But the organization of that genetic variation is also important for locally adapted crop varieties.

Local adaptation in crops is a response to all of the selective forces occurring in the population's environment. Selection, the discrimination between individual genetic contributions to the next generation based on those individuals' fitness (Falconer 1989:26), can be a powerful force in changing allelic diversity and shaping population structure. The changes caused by either natural or artificial selection are the product of the intensity of selection, the phenotypic variability of the population for that trait, and the extent to which the trait is genetically determined (heritability; Falconer 1989:192). Folk varieties are similar to modern varieties in that both are primarily the product of many generations of natural selection plus some limited, recent, artificial (human) selection (Frankel and Soulé 1981; Harlan 1992). However, as a result of the way in which they are developed, many modern varieties have been exposed to far less natural selection in their recent history than is the case for most folk varieties.

The effect of over sixty years of natural selection on a highly variable barley population (Allard 1988) provides some valuable insights into the role natural selection might play in the amount and structure of genetic variation in folk varieties. Three findings are of particular importance. First, over time there was a trend toward increased population fitness measured as reproductive output. This phenotypic trend was corroborated by isozyme data showing an increase in frequencies of alleles at highly polymorphic loci that confer local adaptation and frequencies approaching fixation for less polymorphic or nearly monomorphic loci. In this case the increase in environmental adaptation represents a drop in genetic variation at the population level. Second, locally specific, adaptive gene complexes of increasing size were formed that behaved, in terms of selection, as single major gene complexes. It is likely that when artificial selection is combined with natural selection further linkages may be created, potentially resulting in even larger gene complexes for local adaptation. Third, the local adaptation described in the two findings above led to "ecogeographic" differentiation between populations grown at different sites. These same results were observed among cross-pollinating populations, although to a much lesser extent for the second and third findings.

It is the combination of both local adaptation to often difficult growing conditions and the presence of culturally valued traits such as seed or cob color that can make some folk varieties important cultural symbols that contribute to the maintenance of social relations within and between communities (see, for example, Dennis 1987; Richards 1986). For example, this appears to be a likely hypothesis for the persistence of many Hopi Native American maize folk varieties in Hopi agriculture today (Soleri and Cleveland 1993).

Admittedly, the outline given above is an extremely simplistic description of the forces responsible for the diversity and structure of farmer-managed folk variety populations. Some of the other factors involved will be discussed later. Yet we can make the following genetic interpretation of folk variety "conservation" based on even this limited understanding of these agricultural systems. This perspective is concerned with both the genetic content and structure of folk variety populations, and emphasizes qualities such as overall population structure and the presence of genetic diversity affecting local adaptation. In these agricultural systems, genetic diversity is maintained within the currently cultivated populations, not separate from them. Seen from this perspective, in situ "conservation" is not really conservation at all, as it is commonly defined, but rather the ongoing crop genetic component of a larger agricultural system (Friis-Hansen 1993, Soleri and Smith 1995). This interest in "conservation" for meeting current needs and maintaining future adaptive potential is also being discussed for nondomesticated plant species (Hamilton 1994).

For meeting the needs of farmers, ex situ conservation would be far less desirable than would continued maintenance in the local farming system itself. Sampling for ex situ preservation would require a different approach than is taken for conventional goals, especially a much larger sample size than would be taken when concerned solely with allelic diversity rather than population structure (Marshall and Brown 1983). Preservation and regeneration would need to be particularly sensitive to maintaining local adaptation to the "native" growing environment.

These different perspectives reflect a central, although often unstated, division in the discussions of the "conservation" of folk varieties, between those who see the role of this process as being primarily one of conservation (Brush 1991), and those who see it as ultimately being a local development issue (Cleveland et al. 1994; Cooper et al. 1992). The existence of such differences in perspective is not unique to research concerning the conservation of folk varieties. Historians and philosophers of science have pointed out the role of subjectivity in scientific investigation at many levels, acknowledging, for example, that the researcher's perspective or the assumptions of his or her larger social paradigm can affect research design, choice of hypotheses that are to be tested, and in some cases the findings themselves (Gould 1981; Harding 1991). Recognizing the nature of one's subjectivity is a significant step toward reducing confounding issues in research design and execution. For example, delineating the different perspectives on the conservation of folk varieties as clearly as possible enhances researchers' ability to ask fundamental questions that will best improve our understanding of the conservation process, its potential, and its utility to both farmers and conservationists.

Perspectives on the Conservation Process: An Example

The genetic implications of these different perspectives in terms of conservation goals in general and for specific aspects such as the conservation process itself have not been addressed in the past. From the conventional perspective it is assumed that conservation, accomplished almost exclusively ex situ, generally maintains the genetic fidelity of crop genetic resource collections unaffected by genetic changes (NRC 1991:76). From the farmer perspective, a requirement for conservation to meet their needs is that populations remain locally adapted and thus able to produce an acceptable harvest, and be capable of responding to changing local conditions. To understand whether the conservation process is addressing the concerns of either perspective requires first identifying some of the problems that may occur during that process.

Two significant threats to the conservation of crop genetic resources are genetic drift and genetic shift. Genetic drift is the random change in allele frequencies resulting from the sampling process (Falconer 1989:51), and represents a reduction in effective population size and allelic diversity. Genetic drift operating in two or more subpopulations may result in differentiation between those subpopulations; similarly, genetic drift operating in one subpopulation will result in its being different from the larger population from which it was drawn. The extent of the loss of diversity and the effect on population structure will depend on the severity of the population size reduction and the number of generations that reduction persists (Ellstrand and Elam 1993).

Genetic shift, the natural selection that occurs during ex situ seed maintenance and multiplication (Breese 1989), has been a specific concern of crop genetic resource conservationists and formal plant breeders. Selection changes population structure and can reduce allelic diversity. Its specific effect will depend on the heritability of the traits affected by selection, selection intensity and duration, and the population's original diversity and structure (Falconer 1989:192). Although selection is seen as a threat to the ex situ, static preservation of allelic diversity from the conventional perspective of formal breeders and conservationists, folk varieties are produced and maintained by natural and artificial selection, constant features of the in situ environment. For in situ maintenance to support local farming systems, therefore, the key is to keep selection coefficients sufficiently high to maintain local adaptation while still low enough to conserve as much genetic diversity as possible. "The difficulty," as stated by Frankel, "is to find the borderline between adaptive change and genetic erosion" (Frankel 1970:476). Thus genetic drift may occur in situ but

Table 8.1. Contrast of first and most recent regenerations of USDA Hopi kokoma and blue maize varieties conserved ex situ.

	ORTHOGONAL CONTRAST (F VALUE)	
Characteristic	USDA Kokoma 1956 vs. USDA Kokoma 1989	USDA Blue 1954 vs. USDA Blue 1985
Plant height	6.5*	9.22*
Central spike length	0.25	19.55*
Tassel branch region	9.8*	14.76*
Primary tassel branches	39.44*	1.98
First tassel	1.1	5.46*
Second silk	18.63*	0.03
Second tassel	4.66*	0.20
Ear length	5.60*	6.21*
Ear ratio (diameter/length)	0.98	4.60*
Kernel width	2.07	108.40*
Kernel length	18.72*	1.50
Kernel ratio (width/length)	27.10*	46.98*

Source: Soleri and Smith (1995).
*Contrast is significant at 0.05 level.

genetic shift will not. Both genetic drift and shift may occur ex situ during storage as well as regeneration (see, for example, Roos 1984a, 1984b).

An example of how concerns for the conservation process and threats to genetic diversity such as genetic drift and shift may differ based on the conservation perspective is discussed in a comparison of the in and ex situ conservation of two Hopi Native American maize varieties (Soleri and Smith 1995). In that research, data on the morphological and phenological characteristics of two Hopi maize varieties conserved in situ by Hopi farming households, and the same varieties conserved for over thirty-five years ex situ in Ames, Iowa, by the USDA/ARS/NPGS (U.S. Department of Agriculture/Agricultural Research Service/National Plant Germplasm System) were used to test hypotheses regarding change during ex situ conservation of a variety and differences between in and ex situ populations of the same variety.

The first hypothesis regarding change during ex situ conservation was tested by comparing early with later USDA regenerations (seed multiplications)

Table 8.2. Example of first phenological characteristics of populations of USDA Hopi blue maize conserved ex situ.

Population	MEAN ± STANDARD ERROR[a]	
	First Tassel (c.v. = 17.2%)	First Silk (c.v. = 32.6%)
1954 regeneration	3.0 ± 0.2	2.5 ± 0.2
1962 regeneration	3.7 ± 0.1	3.2 ± 0.2
1973 regeneration	3.7 ± 0.1	3.4 ± 0.2
1985 regeneration	3.9 ± 0.1	3.6 ± 0.2

Source: Soleri and Smith (1995).
[a]Scored on a scale of 1–5 (1 = least advanced stage of development; 5 = most advanced stage of development) sixty-four days after planting.

of a variety. For both varieties used in the study, Hopi blue field maize and Hopi kokoma, a purple-kerneled field maize, significant differences between first and most recent regeneration were found for a majority of the traits documented (table 8.1; Soleri and Smith 1995).

These phenotypic differences likely represent changes in allele frequencies and allelic complexes between populations and the possibility of allele loss. An unreversed and linear change in first tassel phenology (timing of male inflorescence development) was observed in the USDA blue maize populations, suggesting that a genetic shift may have occurred (table 8.2). No significant differences were observed among the 1962, 1973, and 1985 USDA blue regenerations for days to first tassel or first silk (female inflorescence). However, the first regeneration (1954) was significantly different from all later ones for both characteristics. Overall, with the amount of data available from that study, it is difficult to determine if those differences are due to genetic shift or genetic drift. However, it is worth noting that in terms of the conventional goal for folk variety conservation, the implied decrease in allelic diversity is a concern.

Because genetic drift is random, rare (presumably nonadaptive) alleles are eliminated sooner than are common (adaptive) ones (Breese 1989:40–41). This may be of particular concern from the conventional perspective, as researchers search collections for rare alleles that confer unique phenotypes such as resistance to specific pests. If the ex situ environment differs from the "native" one, as is typically the case, genetic shift may reduce or eliminate alleles that are common and adaptive in a population that originated elsewhere. Therefore,

Table 8.3. Contrast of Hopi kokoma and blue maize varieties conserved in situ (Hopi) and ex situ (USDA).

	ORTHOGONAL CONTRAST (F VALUE)	
Characteristic	Hopi Kokoma vs. USDA Kokoma 1989	Hopi Blue vs. USDA Blue 1985
Plant height	15.48*	0.18
Central spike length	10.87*	0.69
Total tillers	5.16*	5.60*
First tassel	8.95*	14.93*
Ear diameter	3.87*	2.37
Ear length	6.39*	1.30
Ear ratio	0.60	3.77*
Kernel width	89.81*	0.58
Kernel length	55.45*	6.13*
Kernel ratio	0.15	6.11*

Source: Soleri and Smith (1995).
*Contrast is significant at 0.05 level.

when folk varieties are conserved ex situ, the greatest concern for farmer conservation goals may be the changes resulting from genetic shift as compared with genetic drift. For example, the timing of reproductive development associated with local adaptation, and often expressed as photoperiodicity, can be extremely important for low-input farmers in marginal areas (Cromwell et al. 1993:44–45).

The second hypothesis, that in and ex situ populations of the same variety are significantly different, was supported by the research findings (table 8.3; Soleri and Smith 1995). This was tested by comparing in situ populations of a variety with the most recent ex situ regeneration of that same variety. There are several possible reasons for the differences observed, and the actual cause could not be determined from this study. However, their difference may be due to the difference between in and ex situ environments and raises questions about the relevance of conventionally defined ex situ conservation for meeting farmers' needs.

Discussion of different genetic goals for conservation is just beginning, and ideas need to be developed and tested. However, this issue warrants attention both because the differences between industrial and traditionally based, low-external-input agricultural systems are substantial, and because of the growing interest in low-input agriculture and support for traditional agricultural communities (Ceccarelli et al. 1992b, 1994; NRC 1989). A growing number of organizations are conducting research and development efforts that will have to address this issue (see, for example, CLADES et al. 1994; Hodgkin et al. 1993; Lamola and Bertram 1994). Ideas for developing research questions useful for a farmer-breeder perspective on folk variety maintenance by farmers are the subject of the next section of this paper.

Research Questions for a Farmer-Breeder Perspective

To date, the research conducted on farmer maintenance of their own folk varieties has not addressed the farmer-breeder perspective. Early work centered on this question: Why do folk varieties persist; that is, why are they retained by agricultural households and communities after the introduction of modern varieties? More recently, a number of descriptive studies of in situ conservation, especially on the number of farmer-named varieties of a species, have been conducted, including those on potato in Peru (Brush 1986; Zimmerer and Douches 1991), maize in southern Mexico (Bellon 1991), rice in Sierra Leone (Richards 1986) and northern Thailand (Dennis 1987), and Hopi Native American crops in the southwestern United States (Soleri and Cleveland 1993). All of these studies report the retention of folk varieties by the agricultural communities studied. The extent of this retention appears to vary with a combination of environmental and sociocultural factors that differ with each situation.

These studies have provided useful insights into farmer maintenance and dispelled the assumption (see, for example, Frankel 1970) that folk varieties would be completely abandoned when modern varieties became available. Yet these studies have also raised many questions, and none of them have investigated the topic from a farmer-breeder perspective. For those interested in supporting farmer management of folk varieties for use in their own agricultural systems, and local crop improvement, it will be valuable to understand farmers' selection practices and to characterize the folk variety populations that are the product of that selection.

Discussion in the following sections focuses primarily on cross-pollinating species, although minor revisions would make it relevant for self-pollinating species as well.

Describing Farmer Selection

Just as the effectiveness of local selection is increasingly acknowledged by plant breeders (Ceccarelli et al. 1992a, 1994; Simmonds 1991b), collaborating with existing artificial selection by the farmer-breeders of the local agricultural community is also being recognized as important for developing crop varieties appropriate for local needs (Maurya et al. 1988; Sperling et al. 1993; Women of Sangams Pastapur and Pimbert 1991). Describing farmer selection practices is important for supporting the farmer perspective for two reasons. First, it is these practices that combine with natural selection to create folk variety populations (Harlan 1992). Second, these practices are the starting place for farmer and researcher collaboration to enhance local crop improvement efforts. Collaboration that builds on those practices is critical to ensure that farmers not only participate but also guide research and development.

Farmers' crop selection practices have not been well studied. The work by Richards (1986) with the Mende rice cultivators of Sierra Leone is the best documentation of systematic farmer selection and experimentation with their folk varieties. He reports local classification systems in which farmers divide their seventy named rice varieties into three classes according to length of their growing cycle and suitability for specific soil and water conditions (Richards 1986:134). Small trial plots were used to evaluate new varieties or selections, and farmers conducted germination tests for seed viability. Artificial selection was applied both through roguing in the field and separate conservation of promising new phenotypes (p. 139).

The effect of artificial selection, resulting from farmer practices, will depend on how and when farmers do their selecting, and the genetic structure of the crop populations. Farmer selection is generally assumed to occur as mass selection. That is, based on each plant's phenotypic performance, individual plants are selected to provide seed for the next generation (Simmonds 1979: 135–42). This process may be repeated each generation.

Mass selection, especially practiced over many generations, can be effective and has been the primary basis of crop improvement since domestication. However, the rate of that improvement depends heavily on the heritability of the trait, which must be high, as selection for traits that are not the product of recombination in cross-pollinating species is based solely on maternal genotypes. For these species the heritability is half of what it would be if the paternal genotype were also selected (Simmonds 1979:139). Although heritability will vary among traits, populations, and environments, the highly heterogeneous environments characteristic of most low-input agricultural fields will also make heritability lower than it would be under more uniform conditions. Thus, if for example, there is little genetic variation for a trait being selected for, or the heritability is extremely low, or if the alleles for two traits being selected for are not genetically correlated in a positive way, selection will be ineffective. Farmers also select their seed sources, which can have an effect on their crop populations. Seed acquisition and distribution practices offer clues to the extent of local adaptation, size of the gene pool, and other issues important for farmer management. For all of the reasons described above, understanding the selection process is part of assessing the potential for improving local crop development.

There is little descriptive information regarding selection and seed distribution practices that will affect population structure. To understand farmer maintenance of folk varieties and its genetic consequences, descriptive information is needed about artificial selection that addresses the following questions:

What traits do farmers use to distinguish among their varieties?
What selection criteria are being used?
When does selection occur—before or after fertilization? in the field? at harvest? during processing?
What are farmers' seed/propagule sources?
What are farmers' planting patterns?

Both artificial and natural selection operating on individual genotypes change allelic frequencies and thus the amount and structure of crop genetic diversity within and between folk variety populations and human communities. Characterization of folk variety populations—specifically, understanding the distribution of the diversity that is important for local adaptation and agronomic performance—is the other information useful for contributing to a farmer-breeder perspective on local folk variety management.

Characterizing Folk Variety Populations

Characterizations of folk varieties are limited. Exceptions to this are work by population geneticists, plant systematists, and plant breeders who have evaluated the genetic diversity of field collections of folk varieties and of their nondomesticated relatives. These studies focus on gaining insights to improve collection strategies for ex situ conservation (Brush et al. 1995; Zimmerer and

Douches 1991), the evaluation of the material for use in breeding programs (Chen et al. 1994; Smith et al. 1995; Wilson et al. 1990), or on insights into the evolution and historical geography of the species (Doebley et al. 1983; Garvin and Weeden 1994; Gepts et al. 1992). All of these studies build upon the consensus that in general, folk varieties contain higher levels of both inter- and intravarietal genetic diversity than do modern variety populations of the same species (see, for example, Frankel and Soulé 1981:201).

But characterization of folk variety populations specifically for supporting the work of farmer-breeders remains to be done. The areas important for such characterizations are these varieties' genetic variation and the structure of that variation—that is, how it is organized in genotypes and distributed within and between populations of a variety. In this section we will outline why genetic variation is valuable from a farmer perspective, and briefly discuss the four other factors in addition to selection that affect folk variety population structures: history of the species in the region, population size, gene flow, and geographic distribution. Although many of these areas have rarely been studied in folk varieties, research on nondomesticated plant species provides some insights into their significance.

Genetic variation, or lack of inbreeding, is considered an important measure of viability and stability of wild plant populations (Ellstrand and Elam 1993; Huenneke 1991). Exceptions to this may be found in some self-pollinating species or species with a long history of small population size, in which frequencies of deleterious genotypes and alleles have presumably been reduced by selection. Similarly, and as mentioned earlier, genetic variation can support a crop variety's ability to survive biotic stresses such as pests and pathogens in their growing environment (Gould 1988, Simmonds 1991a), and is seen as a source of relative yield stability for folk varieties grown in low-input agricultural systems (Ceccarelli et al. 1992b; Cleveland et al. 1994). Genetic variation is important for assessing the potential for microevolutionary change in the future, determining conservation or management strategies, and in the case of folk varieties, aiding local crop improvement efforts. As mentioned, measures of heritability will also be important for determining effective selection criteria for local improvement.

The distribution and structure of intravarietal diversity within and between communities is important because it reflects the consequences of folk variety management, including the effects of natural selection. For example, comparisons of varietal population structure can be made between households or communities. The household is often seen as the primary production and consumption unit and therefore a critical decision making body whose actions affect its environment and resources (Netting 1993). There is reason to believe that this includes the persistence of folk varieties and how they are structured genetically (Bellon and Taylor 1993; Richards 1986; Soleri and Cleveland 1993). There is evidence suggesting that in some cases varietal management and conservation may also operate at a community level (Dennis 1987).

However, investigations of genetic diversity and population structure are typically conducted at the level of the individual accession or population and are concerned with describing overall regional diversity or comparing diversity between regions. Variation at the level of the populations being grown by individual households or communities is rarely the focus of study. Exceptions are two studies of isozyme variation of Andean potato folk varieties (Brush et al. 1995; Zimmerer and Douches 1991). Both those investigations found the majority of allelic diversity to occur within populations of a variety, while genotypic diversity was greatest between populations of the same variety grown by different farmers. The high intrapopulation allelic diversity is thought to be the result of both specific growing environments and management practices that allow recombination to occur in this predominantly vegetatively propagated crop, and maintain its products in the local population of a variety. Genotypic diversity between populations may represent the natural selection forces of different growing environments that favor particular allelic complexes over others (Allard 1988).

Still, as with investigations of intervarietal diversity, these studies focused on evaluation of the populations for determining sampling strategy for ex situ collections. Thus their choice of how to measure genetic variation and their discussion of its significance are based on the conventional approach to folk variety conservation.

Overall, data regarding the simplest descriptions of the structure of subpopulations of a folk variety between households or between communities, or comparisons of these for two or more coexisting varieties, have simply not been available (Frankel and Soulé 1981:201). Instead this information must come from the application of theoretical models (see, for example, Marshall and Brown 1975, 1983) or from research in the population genetics and conservation biology of nondomesticated species that in some cases parallel the situation with domesticates, although that too is very limited (Millar and Libby 1991).

In addition to selection, theory and the available empirical evidence indicate that several other factors will affect the structure of folk variety populations in cross-pollinating species. The evolutionary/cultivation/utilization history of the species in the region of interest provides a theoretical baseline for the relative amount of genetic variation that might be expected (Karron 1991). For example, the center of origin of pearl millet is believed to have been Saharan or sub-Saharan Africa, especially West Africa, with the Sahelian region recognized as the major center of diversity for that species (Brunken et al. 1977). Pearl millet is thought to have reached the coast of what is today western India in 2900 B.P. Given this history, it is likely that the Indian pearl millet gene pool has a reduced level of genetic variation relative to the original African gene pool (founder effect). The extent of this reduction and its mitigation since founding depend on a number of factors.

First, the amount of seed introduced and the variability of the population or populations from which it was taken are unknown. Second, the number and size of subsequent introductions are also unknown. Third, during the nearly three millennia since introduction, the genetic variation of those original founding populations has most likely been augmented by mutation and altered through selection. Using the commonly cited mutation rate of 10^{-5} loci/generation (see, for example, Falconer 1989), assuming one generation/year indicates that almost 3 percent of loci may have experienced mutation since the introduction of pearl millet into Asia.

Reduced population size can have a significant effect on population diversity and structure (Ellstrand and Elam 1993, Menges 1991). Genetic drift results in a small effective population size (N_e) and can occur in an agricultural system when social or biological factors in that system result in the random loss of genetic resources. Genetic drift represents the loss of genetic diversity, specifically in the form of allelic polymorphisms, and over the long term, loss of heterozygosity. Where genetic drift occurs in a population divided into smaller subpopulations, genetic variation is also reorganized, as it is dispersed between those increasingly different subpopulations.

Gene flow, the movement of genetic material between populations, can have a significant effect on plant population structure (Ellstrand and Elam 1993). Gene flow in agricultural systems occurs not only through the dispersion of pollen and seeds by natural agents but also through human-managed seed distribution that can have a large effect on how genetic variation is partitioned among populations of a variety between households, communities, or regions. Ellstrand and Elam (1993:230) found gene flow of over half of rare plant species they surveyed in California to be sufficient to effectively homogenize allele frequencies between populations of a species, and cited gene flow as an important organizer of genetic variation in those species. In the case of folk varieties of a crop species, gene flow could affect the organization of genetic variation within a variety as well as between varieties (varietal purity). Gene flow between local folk varieties and between a folk variety and a cultivar developed outside the area are possibilities for cross-pollinating crop species. The first case will affect the composition of the varieties and be affected by the number of varieties being grown, and farmers' criteria for varietal purity and methods for maintaining this.

Gene flow between local folk varieties and a relatively recent, conspecific introduction may be a valuable source of additional diversity, or it may have consequences for the adaptation of the folk variety populations that are contaminated. How then is the possibility of gene flow between local folk varieties and nonlocal introductions treated in situ? Although this was not addressed directly in that study, it appears that maize farmers in Chiapas, Mexico, did not prevent the mixing of local folk varieties and both open-pollinated and hybrid nonlocal introductions (Bellon 1991). On the other hand, numerous Hopi maize farmers expressed concern regarding the possibility of "contamination" of their Hopi varieties by nonlocal introductions (Soleri and Cleveland, unpublished notes). However, whether this concern is effectively addressed in crop management is not clear.

Gene flow of "inappropriate" alleles may decrease the local adaptation of a population (Ellstrand and Elam 1993:231). This reduction in fitness will occur if the immigration of nonlocal alleles exceeds selection against those alleles. In the case of folk varieties this means that the assumption that genetic diversity has been augmented by the addition of modern varieties to traditional farmers' crop repertoires (see, for example, Brush 1991; Dennis 1987:266) should be reconsidered. While such gene flow may be a "beneficial" source of genetic variation, this is not necessarily so. Indeed, Angermeier (1994) argues that biodiversity as a goal must not always be equated with absolute genetic diversity because there are inherent features of "native diversity" that make it superior to exotic diversity in terms of societal value and evolutionary potential. To the extent that those features contribute to cultural preferences, local adaptation, and evolutionary potential, the same could be postulated about gene flow from recently introduced cultivars to folk varieties.

Finally, geographic distribution can also influence genetic variation with widespread populations tending to have greater genetic variation than do those with a narrower distribution (Ellstrand and Elam 1993:219). This may also be reflected in differences between farmer-named varieties that are regionally common on the one hand, and those confined to a specific location on the other.

Characterization of folk varieties to obtain information more relevant to the needs of farmer-breeders could include evaluation of traits and genotypes important for local adaptation and agronomic performance, especially those identified by local farmers; statistically significant comparisons of folk variety subpopulations between households and between communities; comparisons of groups of traits, or groups of genotypes (depending on the mating system) of subpopulations of the variety. Examples of questions that might be asked are:

Do populations of the same folk variety differ within a community? between communities in a region?
Do two or more folk varieties differ from one another within a community? between communities?
Is genetic variation favored by farmer selection criteria?
What is the heritability of traits favored by farmer selection criteria?

Conclusion

Being aware of the assumptions underlying our research approach will enable those working with the conservation of folk crop varieties to design and conduct research more effectively and to provide information and methods that are relevant to the people who are using those resources. To address the needs of traditionally based, low-external-input agricultural communities requires taking a different perspective on folk variety conservation than is usually assumed. This farmer-breeder perspective recognizes that the local varieties and the genetic diversity they represent are an important part of the current farming system. Folk varieties are not a genetic resource that is being "conserved" in the conventional sense for future use, but rather the actively managed genetic component of the system that is constantly being used and changed.

From this new perspective, it is possible to return to basic concepts from fields such as population genetics and conservation biology and determine their meaning for understanding and asking questions of relevance to farmer-breeders. An emphasis on locally focused research and on farmer participation and collaboration favors the development of relatively rapid, inexpensive, and accessible research methods. Perhaps by approaching local crop improvement as a collaboration between farmers and outside researchers, both of whom manage crop populations, though in different ways, farmer knowledge may be more widely recognized, breeding efforts made more relevant and useful, and local control and continuation of the improvement process become more likely.

References

Allard, R. W. 1988. "Genetic Changes Associated with the Evolution of Adaptedness in Cultivated Plants and Their Wild Progenitors." *Journal of Heredity* 79:225–38.

Angermeier, Paul L. 1994. "Does Biodiversity Include Artificial Diversity?" *Conservation Biology* 8:600–602.

Bellon, Mauricio R. 1991. "The Ethnoecology of Maize Variety Management: A Case Study from Mexico." *Human Ecology* 19:389–418.

Bellon, Mauricio R., and Edward J. Taylor. 1993. "Farmer Soil Taxonomy and Technology Adoption." *Economic Development and Cultural Change* 41:764–86.

Breese, E. L. 1989. *Regeneration and Multiplication of Germplasm Resources in Seed Genebanks: The Scientific Background.* Rome: International Board for Plant Genetic Resources (IBPGR).

Brunken, Jere, J. M. J. De Wet, and J. R. Harlan. 1977. "The Morphology and Domestication of Pearl Millet." *Economic Botany* 31:163–74.

Brush, Stephen B. 1986. "Genetic Diversity and Conservation in Traditional Farming Systems." *Journal of Ethnobiology* 6:151–67.

———. 1991. "A Farmer-Based Approach to Conserving Crop Germplasm." *Economic Botany* 45:153–65.

Brush, Stephen B., Rick Kesseli, Ramiro Ortega, Pedro Cisneros, Karl Zimmerer, and Carlos Quiros. 1995. "Potato Diversity in the Andean Center of Crop Domestication." *Conservation Biology* 9:1189–98.

Ceccarelli, Salvatore, Stefania Grando, and John Hamblin. 1992a. "Relationship between Barley Grain Yield Measured in Low- and High-Yielding Environments." *Euphytica* 64:49–58.

Ceccarelli, Salvatore, J. Valkoun, W. Erskine, S. Weigand, R. Miller, and J. A. G. Van Leur. 1992b. "Plant Genetic Resources and Plant Improvement as Tools to Develop Sustainable Agriculture." *Experimental Agriculture* 28:89–98.

Ceccarelli, Salvatore, W. Erskine, J. Hamblin, and S. Grando. 1994. "Genotype by Environment Interaction and International Breeding Programmes." *Experimental Agriculture* 30:177–87.

Chen, H. B., J. M. Martin, M. Lavin, and L. E. Talbot. 1994. "Genetic Diversity in Hard Red Spring Wheat Based on Sequence-Tagged-Site PCR Markers." *Crop Science* 34:1628–32.

CLADES, COMMUTECH, CPRO-DLO, GRAIN, NORAGRIC, PGRC/E, RAFI, and SEARICE. 1994. "Community Biodiversity Development and Conservation Programme." Proposal to DGIS, IDRC, and SIDA for Implementation Phase I-1994–1997. Wageningen, The Netherlands, and Santiago, Chile: Centre for Genetic Resources, and Centro de Educación y Tecnología.

Cleveland, David A. (n.d.). "Farmer and Formal Plant Breeders: The Anthropology of Genotype-by-environment Interaction in Crop Improvement." Manuscript under review.

Cleveland, David A., Daniela Soleri, and Steven E. Smith. 1994. "Do Folk Crop Varieties Have a Role in Sustainable Agriculture?" *BioScience* 44:740–51.

Cooper, David, Henk Hobbelink, and Renée Vellvé. 1992. "Why Farmer-Based Conservation and Improvement of Plant Genetic Resources?" Pp. 1–16 in *Growing Diversity: Genetic Resources and Local Food Security*. D. Cooper, R. Vellvé, and H. Hobbleink, eds. London: Intermediate Technology Publications.

Cox, T. S. 1991. "The Contribution of Introduced Germplasm to the Development of US Wheat Cultivars." Pp. 25–47 in *Use of Plant Introductions in Cultivar Development*. Part 1. H. L. Shands and L. E. Wiesner, eds. Madison, Wisc.: Crop Science Society of America.

Cromwell, Elizabeth, Steve Wiggins, and Sondra Wentzel. 1993. *Sowing beyond the State*. London: Overseas Development Institute.

Dennis, John Value Jr. 1987. "Farmer Management of Rice Variety Diversity in Northern Thailand." Ph.D. diss., Cornell University.

Doebley, J. F., M. M. Goodman, and C. W. Stuber. 1983. "Isozyme Variation in Maize from the Southwestern United States: Taxonomic and Anthropological Implications." *Maydica* 28:97–120.

Ellstrand, Norman C., and Diane R. Elam. 1993. "Population Genetic Consequences of Small Population Size: Implications for Plant Conservation." *Annual Review of Ecology and Systematics* 24:217–42.

Falconer, D. S. 1989. *Introduction to Quantitative Genetics*. 3d ed. Essex, United Kingdom: Longman Scientific & Technical.

Finlay, K. W., and G. N. Wilkinson. 1963. "The Analysis of Adaptation in a Plant-Breeding Programme." *Australian Journal of Agricultural Research* 14:742–54.

Frankel, O. H. 1970. "Genetic Conservation in Perspective." Pp. 469–89 in *Genetic Resources in Plants: Their Exploration and Conservation*. O. H. Frankel and E. Bennett, eds. Oxford: Blackwell Scientific Publications.

Frankel, Otto H., and Michael E. Soulé. 1981. *Conservation and Evolution*. Cambridge: Cambridge University Press.

Friis-Hansen, Esbern. 1993. "Conceptualizing *In Situ* Conservation of Landraces: The Role of IBPGR." Review paper produced by author as consultant to IBPGR, March–April 1993.

Garvin, D. F., and N. F. Weeden. 1994. "Isozyme Evidence Supporting a Single Geographic Origin for Domesticated Tepary Bean." *Crop Science* 34:1390–95.

Gepts, Paul, T. Stockton, and G. Sonnante. 1992. "Use of Hypervariable Markers in Genetic Diversity Studies—Proceedings of the Symposium on Applications of RAPD Technology to Plant Breeding." Joint Plant Breeding Symposia Series. Crop Science Society of America/American Society for Horticultural Science/American Genetic Association.

Gould, Fred. 1988. "Evolutionary Biology and Genetically Engineered Crops." *BioScience* 38:26–33.

Gould, Stephen Jay. 1981. *The Mismeasure of Man.* New York: W. W. Norton & Company.

Hamilton, Matthew B. 1994. "*Ex Situ* Conservation of Wild Plant Species: Time to Reassess the Genetic Assumptions and Implications of Seed Banks." *Conservation Biology* 8(1):39–49.

Harding, Sandra. 1991. *Whose Science? Whose Knowledge?* Ithaca, N.Y.: Cornell University Press.

Harlan, Jack R. 1992. *Crops and Man.* 2d ed. Madison, Wisc.: American Society of Agronomy, and Crop Science Society of America.

Hodgkin, Toby, V. Ramanatha Rao, and K. Riley. 1993. "Current Issues in Conserving Crop Landraces *In Situ.*" Paper presented at the On-Farm Conservation Workshop, December 6–8, 1993, Bogor, Indonesia.

Huenneke, Laura Foster. 1991. "Ecological Implications of Genetic Variation in Plant Populations." Pp. 31–44 in *Genetics and Conservation of Rare Plants.* D. A. Falk and K. E. Holsinger, eds. New York: Oxford University Press.

Karron, Jeffrey D. 1991. "Patterns of Genetic Variation and Breeding Systems in Rare Plant Species." Pp. 87–98 in *Genetics and Conservation of Rare Plants.* D. A. Falk and K. E. Holsinger, eds. New York: Oxford University Press.

Lamola, Leanna M., and Robert B. Bertram. 1994. "Experts Gather in Mexico to Seek New Strategies in Preserving Agrobiodiversity." *Diversity* 10(3):15–17.

Marshall, D. R., and A. H. D. Brown. 1975. "Optimum Sampling Strategies in Genetic Conservation." Pp. 53–80 in *Crop Genetic Resources for Today and Tomorrow.* O. H. Frankel and J. G. Hawkes, eds. Cambridge: Cambridge University Press.

———. 1983. "Theory of Forage Plant Collection." Pp. 135–48 in *Genetic Resources of Forage Plants.* J. G. McIvor and R. A. Bray, eds. Melbourne, Australia: CSIRO.

Maurya, D. M., A. Bottrall, and J. Farrington. 1988. "Improved Livelihoods, Genetic Diversity and Farmer Participation: A Strategy for Rice Breeding in Rainfed Areas of India." *Experimental Agriculture* 24:311–20.

Menges, Eric S. 1991. "The Application of Minimum Viable Population Theory to Plants." Pp. 45–61 in *Genetics and Conservation of Rare Plants.* D. A. Falk and K. E. Holsinger, eds. New York: Oxford University Press.

Millar, Constance I., and William J. Libby. 1991. "Strategies for Conserving Clinal, Ecotypic, and Disjunct Population Diversity in Widespread Species." Pp. 149–70 in *Genetics and Conservation of Rare Plants.* D. A. Falk, and K. E. Holsinger, eds. Oxford: Oxford University Press.

Netting, Robert McC. 1993. *Smallholders, Householders: Farm Families and the Ecology of Intensive, Sustainable Agriculture.* Stanford: Stanford University Press.

NRC (National Research Council). 1989. *Alternative Agriculture.* Washington, D.C.: National Academy Press.

———. 1991. *Managing Global Genetic Resources: The U.S. National Plant Germplasm System.* Washington, D.C.: National Academy Press.

Plucknett, Donald L., Nigel J. H. Smith, J. T. Williams, and N. Murthi Anishetty. 1987. *Gene Banks and the World's Food.* Princeton: Princeton University Press.

Richards, Paul. 1986. *Coping with Hunger.* London: Allen and Unwin.
Roos, Eric E. 1984a. "Genetic Shifts in Mixed Bean Populations. I. Storage Effects." *Crop Science* 24:240–44.
———. 1984b. "Genetic Shifts in Mixed Bean Populations. II. Effects of Regeneration." *Crop Science* 24:711–15.
Simmonds, N. W. 1979. *Principles of Crop Improvement.* London: Longman Group Ltd.
———. 1991a. "Genetics of Horizontal Resistance to Diseases of Crops." *Biological Review* 66:189–241.
———. 1991b. "Selection for Local Adaptation in a Plant Breeding Programme." *Theoretical and Applied Genetics* 82:363–67.
Smith, S. E., L. Guarino, A. Al-Doss, and D. M. Conta. 1995. "Morphological and Agronomic Affinities among Middle Eastern Alfalfas—Accessions from Oman and Yemen." *Crop Science* 35:1188–94.
Soleri, Daniela, and David A. Cleveland. (n.d.). Unpublished field notes.
———. 1993. "Hopi Crop Diversity and Change." *Journal of Ethnobiology* 13:203–31.
Soleri, Daniela, and Steven E. Smith. 1995. "Morphological and Phenological Comparisons of Two Hopi Maize Varieties Conserved In Situ and Ex Situ." *Economic Botany* 49:56–77.
Sperling, Louise, Michael E. Loevinsohn, and Beatrice Ntabomvura. 1993. "Rethinking the Farmer's Role in Plant Breeding: Local Bean Experts and On-Station Selection in Rwanda." *Experimental Agriculture* 29:509–19.
Walsh, John. 1981. "Genetic Vulnerability Down on the Farm." *Science* 214:161–64.
Wilson, J. P., G. W. Burton, J. D. Zongo, and I. O. Dicko. 1990. "Diversity among Pearl Millet Landraces Collected in Central Burkina Faso." *Crop Science* 30: 40–43.
Women of Sangams Pastapur, Medak, Andhra Pradesh, and Michel Pimbert. 1991. "Farmer Participation in On-Farm Varietal Trials: Multilocational Testing under Resource-Poor Conditions." *Rapid Rural Appraisal (RRA) Notes* Number 10:3–8.
Zimmerer, Karl S., and David S. Douches. 1991. "Geographical Approaches to Crop Conservation: The Partitioning of Genetic Diversity in Andean Potatoes." *Economic Botany* 45:176–89.

PART III

Negotiating the Commons

Ethnoecology Makes a Difference

9

Plant Constituents and the Nutrition and Health of Indigenous Peoples

TIMOTHY JOHNS

Indigenous people are vulnerable to breakdown both of their traditional sociocultural systems and of their traditional ecology. Indigenous societies define themselves in relation to the resources they exploit and in relation to people. The negative impacts of modernization on nutrition and health can be seen as a result of disruptions in both relationships and in the close interconnection between the two.

Over time, finely tuned systems have developed to ensure human survival in habitats with very specific environmental conditions and often limited resources. In environments like the Arctic or the Andean altiplano, successful subsistence systems have developed that draw on locally adapted plants and animals. The savannas of East Africa are clearly well adapted to support large populations of herbivorous ungulates and associated carnivores, and pastoralists such as the Maasai have developed a system compatible with the natural model, in which their animals consume the grass and they depend on the animals for their own sustenance. While these larger ecological dynamics may be fairly obvious, less so are the ways in which traditional systems provide for more subtle needs (such as micronutrients) and through which they contribute to health.

Essential to the survival of indigenous societies is the control over traditional lands, or the commons. Resource procurement to meet basic needs — not the least of which are health and nutrition — provides the best rationale for retaining this control. Additionally, the traditional use of sensitive environments by indigenous peoples is usually more compatible with the long-term conservation of commons resources than are the alternatives. As people are forced to adapt to reduced resource bases, alternative forms of resource utilization, such as managed extraction, cultivation, and agroforestry, may best meet both human needs and the need for conservation.

From the perspective of the modern world, the limitations of particular ways of life may stand out in the face of moves toward economic development. Modernization and associated technology lead to new options and a relaxation of the tight relationship defining how humans utilize the resources in their environment. In many situations, change is thrust upon indigenous peoples, and the adverse effects of resource loss and food insecurity are obvious. However, even where change is embraced, the possible benefits of this modernization process can be offset by the loss of some key component in the traditional way of life. Seldom do we understand the dynamics of traditional systems sufficiently well to predict what the results of such change might be.

Nutrition and the health of indigenous peoples have an ecological basis. The loss or destruction of land, along with reduced access to resources and economic impoverishment, are often at the root of the malnutrition and disease that afflict people when their traditional subsistence economy breaks down. Food insecurity and inadequate intake of essential nutrients, as well as lack of facilities for sanitation or health care, are fundamental problems of poor people in developing countries, and are the focus of major efforts in international health. The road from a disrupted state of ecological and social integrity to one of economic impoverishment associated with dietary inadequacy is familiar. However, the particular problems that afflict indigenous peoples in the process of change to a modern lifestyle, even one of affluence, indicate that the manner in which people successfully survive in a particular environment may involve more subtle processes than is often recognized.

This chapter will examine traditional subsistence systems from a human physiological perspective. Examples of the biological and cultural means by which humans meet their nutritional needs will be considered as a backdrop for understanding the consequences for health and nutritional status when the integrity of traditional systems is destroyed. Humans must deal with toxic chemicals in their environment but also derive benefits from the ingestion of chemicals other than nutrients; the potentially adverse effects of alterations in the way traditional communities interact with these chemicals in their food and elsewhere will be considered in relation to epidemiological data on both environmental toxicity and the incidence of diseases with a dietary basis.

Biological and Cultural Adaptations and Nutrition

The impressive adaptation of human populations to a breadth of global habitats draws on both genetic variation and cultural innovation. The need to procure essential nutrients from diverse sources can account for various enzyme polymorphisms among human populations. The best known example is that of the lactase enzyme, which is deficient among adult members of groups without dairy animals. Likewise, some populations have low activity of the enzyme aldehyde dehydrogenase and are thus more sensitive to ethanol; this could reflect differences in their exposure over many generations to fruit and fermented food.

Exposure over time to particular natural toxins is likely responsible for differences in the activity of liver enzymes involved in detoxication. For example, ethnic differences in the function of oxidative enzymes among human populations result in different rates of metabolism of drugs and toxic compounds (Johns 1990). Similar pressures may contribute to diversity in thresholds for the taste of certain bitter alkaloids and for the capacity to taste compounds such as phenylthiocarbamide (PTC; Johns 1990).

Just as relevant to understanding differences in how indigenous populations meet their subsistence needs in particular environments is cultural diversity in the manner in which humans obtain, prepare, and consume food and deal with other components of the environment. The manner in which traditional preparation of maize (*Zea mays* L.) by Amerindians releases niacin (Carpenter 1981) or the numerous ways in which people process plants to remove toxic phytochemicals (Johns and Kubo 1988) are examples of how people manipulate the chemical constituents of food.

While obtaining essential nutrients and avoiding toxins are fundamental aspects of human physiological adaptation, the important role of nonnutritive plant constituents consumed routinely as normal mediators of health has more recently been recognized (Johns and Chapman 1995). The importance of phytochemicals as the active principles in herbal medicines is well known, but for human populations engaged in traditional forms of subsistence the ingestion of a range of chemicals from plant foods, condiments, beverages, and medicines may be part of their normal ecology. Human interaction with environmental compounds is complex. For populations who live in a closely defined relationship with the natural world, the disruption of the inherent homeostasis in their way of life can be harmful. Indigenous peoples undergoing a process of change, in addition to being susceptible to malnutrition and increased incidence of infectious disease, seem particularly vulnerable to diseases of affluence, perhaps a reflection of the finely tuned nature of their relationship with phytochemicals and other aspects of the environment.

Epidemiology of Toxicity

The loss of traditional food systems may lead to increased exposure to toxic chemicals from plants and other sources. Relationships to environmental toxins that have been established over time can be disrupted by changes in food processing methods or by the consumption of unfamiliar foods. Impoverishment and famine can lead to the consumption of marginal foods along with increased exposure to the toxins they may contain.

Mycotoxins are the naturally occurring contaminants of food that draw the most widespread global concern. Aflatoxins produced by the mold *Aspergillus flavus* growing on seeds of such crops as maize or peanuts (*Arachis hypogaea* L.) are considered a serious problem in many tropical countries and may account for the high rates of liver cancer among populations in West Africa. Pyrrolizidine alkaloids are among the most toxic natural plant compounds (Uiso and Johns 1996a, 1996b), and accidental contamination of food with these compounds is associated with acute epidemics of liver disease (Culvenor 1983). Pyrrolizidine alkaloid toxicity can have a prolonged latent period, and chronic contamination of food grains by seeds of pyrrolizidine alkaloid–containing genera such as *Crotalaria* and *Senecio* is considered to be a factor in the high incidence of primary liver cancer in South Africa. Konzo disease, which is a result of acute exposure to hydrogen cyanide released by poorly processed cassava (*Manihot esculenta* Crantz) roots, occurs in times of drought when cassava is more available than other crops. Chronic exposure to cyanogenic glycosides and other constituents of foods may contribute to the high rates of goiter in certain areas. Incidence of each of these may increase under stress. Toxicity from cassava consumption is more of a problem in Africa, where it was introduced during the period of European colonization, than in Latin America, where people over time have developed effective ways of processing these roots to minimize any harmful effects.

Many indigenous peoples whose animal-derived traditional diets contain minimal amounts of toxins are finding themselves increasingly affected by chemicals from environmental pollution. The long-term consequences of exposure to heavy metals such as mercury and cadmium or to industrial and agricultural chemicals such as PCBs and toxaphene are of great concern to indigenous communities in the Arctic (Chan et al. 1995; Kuhnlein 1995).

Phytochemicals and Diseases of Energy Metabolism

The refined diets of industrial societies, while often more digestible or palatable, differ from those of preindustrial humans in composition. They may not, moreover, be equatable with health. A growing body of epidemiological data links diets high in plant foods with reduced incidence of chronic degenerative diseases such as non-insulin-dependent diabetes mellitus (NIDDM), cancer, and coronary heart disease (CHD; Campbell and Junshi 1994; Stavric 1994) that are widespread in industrial societies. Incidence of these diseases is in part attributed to consumption of lipids and carbohydrates and to a sedentary lifestyle. Understanding of the relative contribution of dietary fat to chronic degenerative disease is confounded by the fact that diets low in fat are typically high in foods of plant origin.

Increasingly, the role of phytochemicals as factors in these diseases is the subject of investigation, although neither the benefits received from these compounds nor the consequences of their elimination from the diets of indigenous people are known. Traditional diets are typically higher in unpalatable and potentially toxic substances from plants than modern diets (Johns 1990, 1994). Moreover, the importance of herbal medicine as a component of traditional ingestion patterns is often overlooked. For many cultures distinctions between food and medicine are weak or even nonexistent (Etkin and Ross, 1982). Without quantitative data on intake of medicinal plants and herbal beverages, our understanding of the importance of phytochemicals in the physiology and health of peoples engaged in traditional lifestyles remains underestimated.

Protectors against Cancer

While aflatoxins or pyrrolizidine alkaloids may contribute to observed differences in rates of cancer incidence among countries, other naturally occurring chemicals may have positive effects in preventing cancer. For example, colon and rectal cancer is common in Europe and North America but is rare in Africa and Asia (Colditz and Willett 1991). The classic work of Burkitt (1971) in associating low incidence of colon cancer among Ugandans with the high intake of vegetable matter stimulated research on the role of dietary fiber in intestinal function and cancer incidence (Kritchevsky and Klurfeld 1991). Other dietary factors such as the portion of total calories from fat, dietary fiber, and the content of fruits and vegetables (Johns and Chapman 1995), including specific protective vegetables such as crucifers and alliums, are among the many environmental differences that distinguish countries with high and low rates of these and other cancers.

Although the mechanisms by which high levels of dietary fat contribute to the development of certain types of cancer (Carroll 1991) are unclear, peroxidation of lipids and the formation of free radicals may be involved. Damage by free radicals to DNA is important in the etiology of cancer.

Naturally occurring substances have been drawn on for decades as the sources of important cancer treatments (Cordell et al. 1991), although the recognition of phytochemicals as "chemopreventives" has been more recent. Natural antioxidants occur in all higher plants (Pratt 1992), and, among these, the antioxidant vitamins C and E and β-carotene have received the most careful scrutiny as chemopreventers, although other constituents of fruits and vegetables may be making a larger contribution to this effect (Johns and Chapman 1995). A variety of widespread dietary plant flavonoids, isoflavonoids, and hydroxycinnamic acids are antioxidants or inhibit tumor development in experimental animals (Huang and Ferraro 1992). Polyphenolic compounds in green tea *(Camellia sinensis)* have been studied extensively as tumor inhibitors (Huang and Ferraro 1992). Common spices that are known to contain phenolics and monoterpenes that inhibit the initiation and development of tumors in animals include rosemary *(Rosmarinus officinalis)* and turmeric *(Curcuma longa)*; Nakatani 1992; Stavric 1994).

Many traditional diets are high in plant foods, many of which can be concentrated sources of chemopreventive chemicals. For example, the East African diets that formed the basis of the classic studies by Burkitt (1971) are high in leafy vegetables and, in addition to containing fiber, are rich in β-carotene and other antioxidant compounds.

Non-Insulin-Dependent Diabetes Mellitus

Non-insulin-dependent diabetes mellitus can reach epidemic proportions in societies undergoing cultural and dietary change. Indigenous populations such as Polynesians and Amerindians who have given up traditional diets in favor of diets that are more processed and higher in fats and calories have suffered sharp increases in NIDDM rates in the last few decades (Young 1993). Among the Tohono O'odham (Pima and Papago) Indians of the American Southwest, approximately 50 percent of adults are afflicted by diabetes.

NIDDM is primarily a disease of energy balance and is strongly associated with obesity (Young 1993). Treatment of NIDDM involves changes in diet and lifestyle, particularly directed at reducing obesity, and oral hypoglycemic agents. Dietary recommendations include reduction in fat intake and greater intake of dietary fiber and complex carbohydrates. An important goal of treatment is to reduce the characteristically elevated blood glucose levels of diabetics and to reduce the exaggerated postmeal glycemic responses. The extent to which other ingested constituents such as nonnutrients in plant foods mediate the glycemic response or have other effects on the development or severity of NIDDM is poorly known.

Traditional diets of the Tohono O'odham came from subsistence farming, hunting, and gathering. Plant foods consumed in the last century included crops such as maize, wheat (*Triticum* spp.), tepary beans (*Phaseolus acutifolius*), and squash (*Cucurbita* spp.), and gathered species such as cactus buds and fruit, mesquite beans (*Prosopis* spp.), wild berries, and wild greens. Boyce and Swinburn (1993) estimated that 70 to 80 percent of the calories in this traditional diet came from carbohydrate; many of these carbohydrates were high in fiber (Brand et al. 1990).

The actual benefits of adopting traditional diets have been demonstrated among Australian aborigines (O'dea 1984) and Hawaiians (Shintani et al. 1991). The latter study emphasized the consumption of traditional Hawaiian sources of complex carbohydrate. These diets led to improvements in both measures of blood sugars and blood lipids (Shintani et al. 1991).

We have summarized plants used as food, food additives, or beverages from which hypoglycemic activity or antidiabetic agents have been identified (Johns and Chapman 1995). Moreover, hypoglycemic and antihyperglycemic activities have been recorded in assays with numerous plants, many of which are used as traditional herbal treatments of diabetes (Marles and Farnsworth 1994). The primary effect of food and beverages with positive effects on diabetes and glycemic control, whether ingested by persons in industrial or nonindustrial societies, is through reduction of the glycemic response from ingested carbohydrate. A glycemic index of foods is calculated from the area under the curve describing the change in blood glucose level that arises over a fixed time after consumption and is used as an indication of the beneficial effects of various sources of carbohydrates (Trout et al. 1993).

Compounds that are widespread in foods of plant origin such as polyphenols, lectins, fiber, and phytic acid likely contribute to the relatively low glycemic index observed in some foods (Jenkins et al. 1994). Thompson (1988,

1993) has demonstrated the role of these compounds in legumes in reducing insulin response and postmeal glucose levels. The mechanism of action in all of these cases is to slow down the digestion of carbohydrate. Through the binding of carbohydrate or of proteins associated with carbohydrate, by inhibiting the activity of digestive enzymes either directly or by chelating mineral cofactors, food constituents such as phytic acid can affect the rate at which glucose enters the blood. The insoluble fiber of guar gum (Ivorra et al. 1989) decreases nutrient absorption by delaying gastric emptying.

Studies on carbohydrate digestion of aboriginal bushfoods suggests that, in addition to widespread compounds such as polyphenolics and phytates, traditional diets may have more specific constituents offering protection against diabetes (Thorburn et al. 1987a). Consider *Castanospermum australe*, Moreton Bay chestnut, which is the species among those tested by Thorburn et al. (1987a) that most reduced glucose and insulin responses in human subjects. This is the source of castanospermine and related indolizidine alkaloids with strong a-glucosidase-inhibiting activity (Saul et al. 1985). Natural sources including microorganisms, algae, and higher plants have demonstrated the presence of a-glucosidase inhibitors (Johns and Chapman 1995). Acarbose (Bayer Pharmaceutical Co.), an a-glucosidase inhibitor in clinical use, does not affect energy intake; rather, instead of glucose from digested starch being absorbed in the top third of the jejunum, it is absorbed over the full length of the small intestine. Compounds displaying this activity have not been systematically studied in the diet.

The indigenous food and medicinal plants of other communities may contain principles similar to Australian bushfoods (Thorburn et al. 1987a), and broader investigation of the hypoglycemic activity of ingested substances may help explain why traditional indigenous patterns of ingestion appear to confer protection against diabetes.

Protectors against Cardiovascular Disease

The so-called French Paradox posed by the lower rates of atherosclerosis and CHD among some populations of industrialized European countries when compared with others with similarly high intake of saturated fat has been attributed to a lower consumption of whole milk and greater consumption of plant foods and red wine (Artaud-Wild et al. 1993). In relation to the understanding of dietary factors contributing to coronary heart disease, the high-fat diets of Inuit present a classical case by which a dietary component with a protective effect, specifically w-3 fatty acids in fish, was identified (Innis et al. 1988).

Among dietary constituents, carotenoids and vitamin E receive the most attention for having a role in retarding the progression of atherosclerosis by protecting against lipoprotein peroxidation (Esterbauer 1989), particularly of polyunsaturated LDL cholesterol (Abbey et al. 1993). Flavonoids (Stavric and Matula 1992) and other polyphenolics in food and beverages with antioxidant properties may lower cholesterol through this mechanism. The apparent positive effects of red wine on platelet aggregation and lipid metabolism may come from resveratrol, tannins, and other phenolics (Johns and Chapman 1995).

Saponins and phytosterols have been shown to have hypocholesterolemic effects in animals and humans (Fraser 1994; Thompson 1993). They may prevent cholesterol absorption, interfere with its enterohepatic circulation, and increase its fecal excretion. Water-soluble dietary fiber, such as pectin, β-glucan, and certain gums, also lowers plasma cholesterol (Johns and Chapman 1995).

Phytochemicals in the Cholesterol Metabolism of the Maasai

The low incidence of cardiovascular disease among the Maasai of East Africa was studied extensively during the 1960s and 1970s in relation to the high cholesterol- and fat-containing diet of this pastoral population, but it was not satisfactorily explained (Chapman et al. 1997). The Maasai provide a paradox between diet and coronary heart disease comparable to that in Europe, but in a traditional ecological context.

The Maasai maintain a traditional subsistence lifestyle as pastoralists. Like most such populations, their major health problems are related to undernutrition, sanitation, and infectious disease, and they are relatively free of the diseases of industrial societies. The Maasai obtain as much as 66 percent of their calories from fat (Biss et al. 1971), and despite subsistence on a diet high in milk, yogurt, and meat, they are not subject to hypercholesterolemia (Mann et al. 1972).

Several theories, none of which is totally satisfactory, have been presented to account for this paradox (Johns and Chapman 1995). Day et al. (1976) suggested that the low serum cholesterol of rural Maasai might be a result of herb consumption but published nothing subsequently in this regard. Many ancillary reports point out the limited contributions of plants to Maasai diets. Such accounts ignore data on Maasai ethnobotany and are invalid. In fact, the Maasai and other East African ethnic groups routinely ingest wild plant materials as foods, as ingredients of milk and meat-based soups, as masticants, and as herbal medicines.

In light of the numerous reports of roots and barks being regularly added to fat-containing food by the Maasai and neighboring tribes (Johns et al. 1994), we are reconsidering the epidemiological data on Maasai diet and cardiovascular status from the perspective of the hypocholesterolemic and antioxidant activities of nonnutritive plant constituents in diet. Based on a literature survey and an ethnobotanical study, we have tabulated a list of twenty-five species that the Maasai in both Kenya and Tanzania add to food. Maasai usually consume meat with soup, and soups almost always contain additives of roots or bark in levels that make the food bitter.

Although these additives have medicinal properties (Johns et al. 1994), most of the plant materials are not necessarily used in a curative manner; instead, their purported purpose is preventative. They are tonics, mild stimulants, or used for the routine relief of chronic afflictions like rheumatism and the aches and pains of daily life. The effects of these rather nonspecifically used plants and their chemical constituents on dietary ecology may extend beyond the immediately stated roles.

Eighty-two percent (9 out of 11) of the Maasai food additives that we screened contain potentially hypocholesterolemic saponins or phenolics. Using in vitro assays related to the proposed mechanisms of saponin action in removing cholesterol from the body, we showed plants such as *Acacia goetzii, Albizia anthelmintica*, and *Myrsine africana* that contain saponins interact with cholesterol (Chapman et al. 1997).

Likewise, plants ingested in a number of other forms probably contribute considerable amounts of phytochemicals. Those wild food plants, which are known to have medicinal properties, may have a broader physiological role than simply as nutrient sources. Of particular note are chewed gums and resins from several species of which *Commiphora africana* is the most important. Hypolipidemic activity has been extensively studied in resin of an Indian species, *Commiphora mukul* (Satyavati 1988), and a similar effect from chewing African members of the genus is likely. Such activity has been attributed to phytosterols. The Maasai also chew barks and roots as thirst quenchers. Honey beer is the most popular beverage in the area, and this is always made with the addition of roots of *Aloe volkensii*, a member of a medicinally important genus with antidiabetic properties (Ivorra et al. 1989). Based on our initial investigation, we hypothesize that saponins, polyphenols, and phytosterols in foods, medicines, and masticants contribute to the phenomenon of low rate of atherosclerosis and CHD among the Maasai.

The Maasai utilize their natural environment to meet their needs in a manner consistent with the practices of their ancestors. Savanna grasses provide food for the large herds of cattle and goats from which the Maasai derive their staple foods. However, increasing population and the influence of modern technology and economic systems are placing intense pressure on pastoral people. They are in competition with agriculturalists for land and water and find it more and more difficult to maintain the seminomadic way of life. In addition to the difficulty in maintaining large herds and the consequences of having less milk, meat, and blood, there are other potential consequences for the Maasai of the disruption of their relationship with their traditional resources. Maize meal that is either purchased or grown is an increasing part of their diet. Potentially less available are the wild fruits and roots that make up an important part of the diet of children, as well as honey and the gums, masticants, and food additives discussed above. Maasai rely on savanna trees and shrubs for most of their herbal medicines. The importance of these resources is underlined by the fact that the Maasai word for tree and their word for medicine are the same, *ol-chani*.

There have been few studies on the impact of modernization or urbanization on the health of the Maasai (Day et al. 1976), although there is some evidence of increasing undernutrition. Suggestions that the Maasai tolerance for fat has a genetic basis have not been substantiated. In fact the Maasai are genetically heterogeneous, although the existence of polymorphisms in enzymes related to fat metabolism could be important in their dietary ecology. No matter whether the adaptations of the Maasai are biological or cultural, changes in staple foods or in the Maasai patterns of ingestion of nonstaples will likely have an impact on their health.

Conclusions

On a fundamental level, human subsistence is mediated by the need for energy and, secondarily, for protein. Human appetite, food preferences, and consumption patterns are directed by the desire to obtain energy, which in natural environments is often of limited availability. Food insecurity and undernutrition are typically associated with inadequate intake of energy and protein. In preindustrial subsistence systems, if humans are successful in procuring adequate energy, this usually ensures a balanced intake of vitamins, minerals, and nonnutrients such as fiber and phytochemicals. In addition to their nutritional and metabolic functions discussed in this paper, vitamins and phytochemicals can have positive effects on immune function or may have antibiotic effects that are important in controlling parasites and infectious disease (Johns 1990).

Diets of groups living a traditional subsistence lifestyle typically maintain a homeostatic balance from plant and animal foods. The Maasai diet dominated by milk and meat is undoubtedly very different from that of ancestral humans, but through the specific ingestion of particular plant products the Maasai appear to maintain the relationship between energy and plant constituents. The diets of subsistence agriculturalists may be based on only a few staples, which by themselves may not be nutritionally balanced. However, in traditional situations people may draw on wild plants or minor crops as relishes, condiments, and food supplements. Decreased use of these resources is associated with vitamin A and other deficiencies and is undoubtedly caused by a reduced intake of beneficial phytochemicals.

The high incidence of diabetes, obesity, or other conditions among many indigenous peoples who adopt the highly refined and energy-dense diets of industrial societies raises the question as to whether such groups are more vulnerable to dietary change than humans in general, perhaps because of their long-standing relationship with resources in an environment. The environmentally determined polymorphisms discussed above would allow for such as possibility. Moreover, the dramatic rates of diabetes among certain populations led to hypothesis of the existence of a "thrifty-geneotype" (Young 1993). The incidence of diabetes among most indigenous groups in North America is, however, much lower than that of the Pima and Papago and relatively closer to that of the North American population as a whole. For the majority of indigenous groups, disruption of their culturally mediated relationships with resources is probably more at play in making them vulnerable to the process of modernization.

Indigenous people enter into modern systems at a disadvantage. Even in affluent societies, they may be relatively poor and they may not have access to all options through which members of these societies satisfy their needs. From a nutritional perspective, economic constraints may make them dependent on low-quality foods. This can be the case in the Canadian north, where fears of contamination in fish and wildlife may push people to depend on imported foods available in supermarkets, foods that are expensive and sometimes nutritionally inadequate (Kuhnlein 1995).

Many of the options for satisfying one's needs within modern societies may not even be fully understood by indigenous persons. They may have limited access to education, or to a form of education that is compatible with traditional values or knowledge systems. Indigenous people moving into larger societies seldom share fully in the culture of those societies. As they straddle two worlds, neither are they able to fully retain their own culture. If traditional culture is defined by relationships to resources and by relationships to people, then loss of the relationship with resources means a loss of cultural values and of cultural integrity. This can have an adverse effect on social relationships and lead to a loss of identity. The Maasai, for example, are people of cattle. Social roles are determined by the requirements of caring for cattle and preparing their products, and social relationships involve giving and receiving of cattle and cattle products. Without cattle, the Maasai know that they will no longer be Maasai. They do not know, furthermore, who in that situation they would be. Undoubtedly, as individuals they would suffer on many levels.

In the inevitable process of transition that faces people moving from a traditional to a modern way of life, indigenous resources and the knowledge of them can play a central and critical role. From the point of view of the discussion in this paper, such resources are important to the health of people; they are often superior to alternate foods and provide biological benefits that may be unrecognized. In the process, indigenous knowledge can be a focal point for the maintenance of cultural pride and integrity, and the procurement and preparation of indigenous foods can help to enforce and restore social relationships.

Crucial to reducing the impact of integration into modern economic systems is the documentation and scientific validation of traditional knowledge. Indigenous systems need to be understood simultaneously on their own terms and in Western scientific terms. In order for indigenous peoples to adapt to the modern world in a timely and coherent way, they need to be participants in research. In addition to greater access to basic education, more indigenous people need assistance and training in modern science and technology in order to allow them to transform the knowledge that is their heritage into a form that has the potential to address the changes that the world of technology imposes (Johns et al. 1994).

References

Abbey, M., P. J. Nestel, and P. A. Baghurst. 1993. "Antioxidant Vitamins and Low-densitylipoprotein Oxidation." *American Journal of Clinical Nutrition*. 58:525–32.

Artaud-Wild, S. M., S. L. Connor, G. Sexton, and W. E. Connor. 1993. "Differences in Coronary Mortality Can Be Explained by Differences in Cholesterol and Saturated Fat Intakes in 40 Countries but Not in France and Finland." *Circulation* 88:2771–79.

Bang, H. O., J. Dyerberg, and H. M. Sinclair. 1980. "The Composition of the Eskimo Food in Northwestern Greenland." *American Journal of Clinical Nutrition* 33: 2657–61.

Beecher, C. W. W. 1994. "Cancer Preventive Properties of Varieties of *Brassica oleracea*: A Review." *American Journal of Clinical Nutrition* 59:1166S–70S.

Bennett, P. 1983. "Diabetes in Developing Countries and Unusual Populations." Pp. 43–57 in *Diabetes in Epidemiological Perspective*. U. Mann, K. Pyorala, and A. Teuscher, eds. Edinburgh: Churchill Livingstone.

Birt, D. F., and E. Bresnick. 1991. "Chemoprevention and Nonnutrient Components of Vegetables and Fruits." Pp. 221–60 in *Cancer and Nutrition*. R. B. Alfin-Slater and D. Kritchevsky, eds. New York: Plenum Press.

Biss, K., K. J. Ho, B. Mikkelson, L. Lewis, and C. B. Taylor. 1971. "Some Unique Biological Characteristics of the Masai of East Africa." *New England Journal of Medicine* 284:694–99.

Bjorck, I., and N. G. Asp. 1994. "Controlling the Nutritional Properties of Starch in Foods—A Challenge to the Food Industry." *Trends in Food Science and Technology* 5:213–17.

Boyce, V. L., and B. A. Swinburn. 1993. "The Traditional Pima Indian Diet—Composition and Adaptation for Use in a Dietary Intervention Study." *Diabetes Care* 16:369–71.

Brand, J. C., B. J. Snow, G. P. Nabhan, and A. S. Truswell. 1990. "Plasma Glucose and Insulin Responses to Traditional Pima Indian Meals." *American Journal of Clinical Nutrition* 51:416–20.

Burkitt, D. P. 1971. "Epidemiology of Cancer of the Colon and Rectum." *Cancer* 28: 3–13.

Campbell, C., and C. Junshi. 1994. "Diet and Chronic Degenerative Diseases: Perspectives from China." *American Journal of Clinical Nutrition* 59:1153–61S.

Carpenter, K. J. 1981. "Effects of Different Methods of Processing Maize on Its Pellagragenic Activity." *Federation Proceedings* 40:1531–35.

Carroll, K. K. 1991. "Dietary Fats and Cancer." *American Journal of Clinical Nutrition* 53:1064S–67S.

Chan, H. M., C. Kim, K. Khoday, O. Receveur, and H. V. Kuhnlein. 1995. "Assessment of Dietary Exposure to Trace Metals in Baffin Inuit Food." *Environmental Health Perspectives* 103:740–46.

Chapman, L., T. Johns, and R. L. A. Mahunnah. 1997. "Saponin-like *In Vitro* Characteristics of Extracts from Selected Non-nutrient Wild Plant Food Additives Used by Maasai in Meat and Milk Based Soups." *Ecology of Food and Nutrition* 36:1–22.

Colditz, G. A., and W. C. Willett. 1991. "Epidemiologic Approaches to the Study of

Diet and Cancer." Pp. 51–67 in *Cancer and Nutrition*. R. B. Alfin-Slater and D. Kritchevsky, eds. New York: Plenum Press.
Cordell, G. A., C. W. W. Beecher, and J. M. Pezzuto. 1991. "Can Ethnopharmacology Contribute to the Development of New Anticancer Drugs?" *Journal of Ethnopharmacology* 32:117–33.
Culvenor, C. C. J. 1983. "Estimated Intakes of Pyrrolizidine Alkaloids by Humans: A Comparison with Dose Rates Causing Tumors in Rats." *Journal of Toxicology and Environmental Health* 11:625–35.
Day, J., M. Carruthers, A. Bailey, and D. Robinson. 1976. "Anthropometric, Physiological and Biochemical Differences between Urban and Rural Maasai." *Atherosclerosis* 23:357–61.
Diamond, J. M. 1992. "Human Evolution—Diabetes Running Wild." *Nature* 357: 362–63.
Esterbauer, H. 1989. "Role of Vitamin E and Carotenoids in Preventing Oxidation of Low Density Lipoproteins." *Annals of the New York Academy of Science* 570: 254–67.
Etkin, N. L., and P. J. Ross. 1982. "Food as Medicine and Medicine as Food." *Social Science and Medicine* 16:1559–73.
Flatz, G., and H. Rotthautwe. 1973. "Lactose, Nutrition and Natural Selection." *Lancet* 1973:16–17.
Fraser, G. E. 1994. "Diet and Coronary Heart Disease: Beyond Dietary Fats and Low-density-lipoprotein Cholesterol." *American Journal of Clinical Nutrition* 59: S1117–S23.
Goedde, H. W., and D. P. Agarwa. 1986. "Aldehyde Oxidation: Ethnic Variations in Metabolism and Response." Pp. 113–38 in *Ethnic Differences in Reactions to Drugs and Xenobiotics*. W. Kalaw, H. W. Goedde, and D. P. Agarwal, eds. New York: Alan R. Liss.
Haskell, W. L., G. A. Spiller, C. D. Jensen, B. K. Ellis, and J. E. Gates. 1992. "Role of Water-soluble Dietary Fiber in the Management of Elevated Plasma Cholesterol in Healthy Subjects." *American Journal of Cardiology* 69:433–39.
Higginson, J., and M. J. Sheridan. 1991. "Nutrition and Human Cancer." Pp. 1–50 in *Cancer and Nutrition*. R. B. Alfin-Slater and D. Kritchevsky, eds. New York: Plenum Press.
Huang, M. T., and T. Ferraro. 1992. "Phenolic Compounds in Food and Cancer Prevention." Pp. 8–34 in *Phenolic Compounds in Food and Their Effects on Health. Vol. II: Antioxidants and Cancer Prevention*. M.-T. Huang, C.-T. Ho, and C.-Y. Lee, eds. American Chemical Society, ACS Symposium Series 507, Washington, D.C.
Hvidberg, E. F. 1986. "Ethnic Differences in Phenytoin Kinetics." Pp. 279–88 in *Ethnic Differences in Reactions to Drugs and Xenobiotics*. W. Kalow, H. W. Goedde, and D. P. Agarwal, eds. New York: Alan R. Liss.
IARC. 1987. "IARC Monographs on the Evaluation of the Carcinogenic Risk of Chemicals to Humans: Aflatoxins." Supp. 7, pp. 83–87. Lyon: International Agency for Research on Cancer.

Innis, S. M., H. V. Kuhnlein, and D. Kinloch. 1988. "The Composition of Red Cell Membrane Phospholipids in Canadian Inuit Consuming a Diet High in Marine Mammals." *Lipids* 23:1064–68.

Ivorra, M. D., M. Paya, and A. Villar. 1989. "A Review of Natural Products and Plants as Potential Antidiabetic Drugs." *Journal of Ethnopharmacology* 27:243–75.

Iwu, M. M., C. O. Okunji, G. O. Ohiaeri, P. Akah, D. Corley, and M. S. Tempesta. 1990. "Hypoglycemic Activity of Disocoretine from Tubers of *Dioscorea dumetorum* in Normal and Alloxan Diabetic Rats." *Planta Medica* 56:264–67.

Jenkins, D. J. A., A. L. Jenkins, T. M. S. Wolever, V. Vuksan, and V. Rao. 1994. "Low Glycemic Index: Lente Carbohydrates and Physiological Effects of Altered Food Frequency." *American Journal of Clinical Nutrition* 59(Supp.):S706–9.

Jensen, C. D., G. A. Spiller, J. E. Gates, A. F. Miller, and J. H. Whittam. 1993. "The Effect of Acacia Gum and Water-soluble Dietary Fiber Mixture on Blood Lipids in Humans." *Journal of the American College of Nutrition* 12:147–54.

Johns, T. 1990. *With Bitter Herbs They Shall Eat It: Chemical Ecology and the Origins of Human Diet and Medicine*. Tucson: University of Arizona Press.

———. 1994. "Defense of Nitrogen-rich Seeds Constrains Selection for Reduced Toxicity during the Domestication of Grain Legumes." Pp. 151–67 in *Advances in Legume Systematics*. Vol. 5: *The Nitrogen Factor*. J. L. Sprent and D. McKey, eds. Kew: Royal Botanic Gardens.

Johns, T., H. M. Chan, O. Receveur, and H. V. Kuhnlein. 1994. "Commentary on the ICN World Declaration on Nutrition: Nutrition and the Environment of Indigenous Peoples." *Ecology of Food and Nutrition* 32:81–87.

Johns, T., and L. Chapman. 1995. "Phytochemicals Ingested in Traditional Diets and Medicines as Modulators of Energy Metabolism." Pp. 161–88 in *Phytochemistry of Medicinal Plants: Recent Advances in Phytochemistry* 29. J. T. Amason, R. Mata, and J. T. Romeo, eds. New York: Plenum Press.

Johns, T., and I. Kubo. 1988. "A Survey of Traditional Methods Employed for the Detoxification of Plant Foods." *Journal of Ethnobiology* 8:81–129.

Johns, T., E. B. Mhoro, and P. Sanaya. 1996. "Food Plants and Masticants of the Baterni of Ngorongoro District, Tanzania." *Economic Botany* 50:115–21.

Johns, T., E. B. Mhoro, P. Sanaya, and E. K. Kimanani. 1995. "Herbal Remedies of the Batemi of Ngorongoro District, Tanzania: A Quantitative Appraisal." *Economic Botany* 48:90–95.

Katz, S. H., and J. I. Schall. 1986. "Favism and Malaria: A Model of Nutrition and Biocultural Evolution." In *Plants in Indigenous Medicine and Diet: Biobehavioral Approaches*. N. L. Etkin, ed. Bedford Hills, N.Y.: Redgrave Publishing Co.

King, H., and M. Rewers. 1993. "Global Estimates for Prevalence of Diabetes Mellitus and Impaired Glucose Tolerance in Adults." *Diabetes Care* 16:157–77.

Kritchevsky, D., and D. M. Klurfeld. 1991. "Dietary Fiber and Cancer." Pp. 127–40 in *Cancer and Nutrition*. R. B. Alfin-Slater and D. Kritchevsky, eds. New York: Plenum Press.

Kuhnlein, H. V. 1995. "Benefits and Risks of Traditional Food of Indigenous Peoples:

Focus on Dietary Intakes of Arctic Men." *Canadian Journal of Physiology and Pharmacology* 73:765–71.

Lean, M. E. J., and J. I. Mann. 1991. "Obesity, Body Fat Distribution and Diet in the Aetiology of Non-insulin-dependent Diabetes Mellitus." Pp. 181–91 in *Textbook of Diabetes*. J. Pickup and G. Williams, eds. Oxford: Blackwell Scientific Publications.

Lieberman, L. S. 1987. "Biocultural Consequences of Animals Versus Plants as Sources of Fats, Proteins and Other Nutrients." Pp. 225–58 in *Food and Evolution: Toward a Theory of Human Food Habits*. M. Harris and E. B. Ross, eds. Philadelphia: Temple University Press.

Mann G., A. Spoerry, M. Gray, and D. Jarashow. 1972. "Atherosclerosis in the Masai." *American Journal of Epidemiology* 95:26–37.

Mann, J., and A. Houston. 1983. "The Aetiology of Non-insulin Dependent Diabetes Mellitus." Pp. 122–64 in *Diabetes in Epidemiological Perspective*. J. I. Mann, K. Pyorala, and A. Teuscher, eds. Edinburgh: Churchill Livingstone.

Marles, R. J., and N. R. Farnsworth. 1994. "Plants as Sources of Antidiabetic Agents." *Economic and Medicinal Plant Research*. 6:149–87.

Nakatani, N. 1992. "Natural Antioxidants from Spices." Pp. 72–86 in *Phenolic Compounds in Food and Their Effects on Health. Vol. II: Antioxidants and Cancer Prevention*. M.-T. Huang, C.-T. Ho, and C.-Y. Lee, eds. American Chemical Society, ACS Symposium Series 507, Washington, D.C.

Nestel, P., and C. Geissler. 1985. "Potential Deficiencies of a Pastoral Diet: A Case Study of the Maasai." *Ecology of Food and Nutrition* 19:1–10.

O'dea, K. 1984. "Marked Improvement in Carbohydrate and Lipid Metabolism in Diabetic Australian Aborigines after Temporary Reversion to Traditional Lifestyle." *Diabetes* 33:596–603.

Pratt, D. E. 1992. "Natural Antioxidants from Plant Material." Pp. 54–71 in *Phenolic Compounds in Food and Their Effects on Health. Vol. II: Antioxidants and Cancer Prevention*. M.-T. Huang, C.-T. Ho, and C.-Y. Lee, eds. American Chemical Society, ACS Symposium Series 507, Washington, D.C.

Satyavati, G. V. 1988. "Gum Guggul *(Commiphora mukul)*—The Success Story of an Ancient Insight Leading to a Modern Discovery." *Indian Journal of Medical Research* 87:327–35.

Saul, R., J. J. Ghidoni, R. J. Molyneux, and A. D. Elbein. 1985. "Castanospermine Inhibits Alpha-Glucosidase Activities and Alters Glycogen Distribution in Animals." *Proceedings of the National Academy of Science* 82:93–97.

Shintani, T. T., C. K. Hughes, S., Beckham, and H. K. O'Connor. 1991. "Obesity and Cardiovascular Risk Intervention through the Ad Libitum Feeding of Traditional Hawaiian Diet." *American Journal of Clinical Nutrition* 53:S1647–51.

Stavric, B. 1994. "Antimutagens and Anticarcinogens in Foods." *Food Chemistry and Toxicology* 32:79–90.

Stavric, B., and T. I. Matula. 1992. "Flavonoids in Foods—Their Significance for Nutrition and Health." Pp. 274–94 in *Lipid- Soluble Antioxidants: Biochemistry and Clinical Applications*. A. S. H. Ong and L. Packer, eds. Basel: Birkhauser Verlag.

Thompson, L. U. 1988. "Antinutrients and Blood Glucose." *Food Technology* 42:123–32.

———. 1993. "Potential Health Benefits and Problems Associated with Antinutrients in Foods." *Food Research International* 26:131–49.

Thorburn, A. W., J. C. Brand, V. Cherikoff, and A. S. Truswell. 1987b. "Lower Postprandial Plasma Glucose and Insulin after Addition of *Acacia coriacea* Flour to Wheat Bread." *Australia and New Zealand Journal of Medicine* 17:24–26.

Thorburn, A. W., J. C. Brand, and A. S. Truswell. 1987a. "Slowly Digested and Adsorbed Carbohydrate in Traditional Bushfoods: A Protective Factor against Diabetes?" *American Journal of Clinical Nutrition* 45:98–106.

Trout, D. L., K. M. Behall, and O. Osilesi. 1993. "Prediction of Glycemic Index in Starchy Foods." *American Journal of Clinical Nutrition* 58:873–78.

Uiso, F. C., and T. Johns. 1996a. "Consumption Patterns and Nutritional Contribution of *Crotalaria brevidens* in Tarime District, Tanzania." *Ecology of Food and Nutrition* 35:59–69.

———. 1996b. "Risk Assessment of the Consumption of a Pyrrolizidine Alkaloid Containing Indigenous Vegetable, *Crotalaria brevidens*." *Ecology of Food and Nutrition* 35:111–19.

Wagner, H., and A. Proksch. 1985. "Immunostimulatory Drugs of Fungi and Higher Plants." Pp. 113–45 in *Economic and Medicinal Plant Research*. Vol. 1. H. Wagner, H. Hikino, and N. R. Farnsworth, eds. London: Academic Press.

Willet, W. C. 1994. "Micronutrients and Cancer Risk." *American Journal of Clinical Nutrition* 59:S1162–65.

Winkler, E. M., and R. R. Sokal. 1987. "A Phenetic Classification of Kenyan Tribes and Subtribes." *Human Biology* 59:121–45.

Woolhouse, N. M. 1986. "The Debrisquine/Sparteine Oxidation Polymorphism: Evidence of Genetic Heterogeneity among Ghanaians." Pp. 39–54 in *Ethnic Differences in Reactions to Drugs and Xenobiotics*. W. Kalow, H. W. Goedde, and D. P. Agarwal, eds. New York: Alan R. Liss.

Young, T. K. 1993. "Diabetes Mellitus among Native Americans in Canada and the United States: An Epidemiological Review." *American Journal of Human Biology* 5:399–413.

10

Sustainable Production and Harvest of Medicinal and Aromatic Herbs in the Sierras de Córdoba Region, Argentina

MARTA LAGROTTERIA
JAMES M. AFFOLTER

"Wildcrafting," the harvest of plants from natural habitats, is one of the primary ways that many medicinal and aromatic species are brought to commercial markets. Some species can be collected in this manner without threatening the survival of natural populations, particularly when collecting techniques are relatively nondestructive. Worldwide, however, many aromatic and medicinal plant species have become endangered or even extinct as a result of overcollection for commercial purposes (Akerele 1991; Cunningham 1991; Riddle 1992: 27–28). Wildcrafting of medicinal and aromatic plants is a conservation and economic issue in the United States, where herbal teas and medicines play a minor yet expanding role in the domestic health care market. In many developing countries, where plant medicines are used on a much larger scale, the pressure on native plant populations in areas considered to be the commons is even more intense (Cunningham 1991; Xiao 1991; Given 1994).

Unlike extracted mineral or petroleum resources, cultivated plants are a renewable resource. Overexploited natural populations can, however, be driven to extinction. Collectors often have a considerable store of traditional and local knowledge to guide them in their search for these species. Yet for commercially wildcrafted plants, local extinction is often preceded by a period in which they become so difficult to locate and collect in sufficient numbers that they can no longer be profitably or sustainably harvested. When commercially valuable plants are wildcrafted from public lands, or when collectors harvest from private land without the owner's permission, many issues are raised concerning ownership, exploitation, and conservation of the commons. The risk of harvesting at nonsustainable rates by individuals motivated largely by self-interest—and, in the case of small-scale collectors, by self-preservation—raises the possibility of a "tragedy" of a scale not too dissimilar from the one witnessed in the case of overgrazed pastoral lands.

This chapter describes an ongoing project in the province of Córdoba, Argentina, to promote the sustainable harvest and horticultural production of native medicinal and aromatic herbs. It illustrates the importance of understanding the local context of livelihood activities in promoting equitable and sustainable use of the natural resources of the commons, as well as some of the obstacles encountered in implementing such strategies.

Diversity and Economic Significance of the Medicinal and Aromatic Flora of Córdoba

The use of herbal teas and medicines is popular in Argentina, and the centrally located province of Córdoba is renowned as a source of medicinal and aromatic species. Three indigenous Native American cultures—Diaguitas, Comechingones, and Sanavirones—contribute to the strong tradition of herbal medicine in the province. These indigenous systems of phytotherapy were enriched with species introduced by the Spanish conquistadors. Unlike the United States, where consumers often view herbal remedies as primitive or ineffectual, many people in Argentina prefer herbal medicines and purchase them in a relatively unprocessed form from street vendors, herb shops, natural food stores, and pharmacies. Individual herbs are typically dried, chopped, and then packaged and sold individually or as mixtures of several species. Consumption is generally in the form of medicinal teas.

Herbs are also used on a large scale in the manufacture of "*yerba mate compuesta*." Mate, the traditional drink of Argentina, is a caffeinated beverage prepared from the leaves of *Ilex paraguariensis* St. Hil., a shrub in the holly family *(Aquifoliaceae)* that is native to Paraguay and adjacent regions of Argentina and Brazil. *Ilex paraguariensis* is cultivated on a large scale, but aromatic herbs collected from the wild are frequently added for flavor. Wildcrafted herbs are also used in Argentina to flavor other beverages and in the production of cosmetics.

Marta Lagrotteria and colleagues began collecting data in 1984 concerning the harvesting and marketing of medicinal and aromatic herbs in Córdoba. The primary study site has been Departamento San Javier, located on the western border of the province. San Javier occupies the slopes and piedmont of the Sierra de Comechingones, which is in turn part of a larger range of mountains, known collectively as the Sierras de Córdoba, that traverses the western portion of the province. San Javier is bordered on the north by Dpto. San Alberto, on the east by Dpto. Calamuchita, and on the south and west by the province of San Luis. Elevation in Dpto. San Javier ranges from 533 meters (Villa Dolores) to 2,790 meters (Cerro Champaquí). The region is semiarid, with median annual precipitation ranging from fifty to eighty centimeters. Streams and rivers descend from Cerro Champaquí and disappear in the alluvial terrain. The soils of the plains are rich in organic sediments and minerals. Although the region is not suitable for cultivation of cereal crops, there are favorable conditions for vineyards and the cultivation of citrus, tobacco, and medicinal and aromatic herbs. The capital city of Córdoba, the political and economic center of the province, is located on the eastern side of the mountains at a distance of approximately 125 kilometers.

Several important centers of collection and primary processing of herbs are found in Dpto. San Javier. In towns such as La Paz, Loma Bola, and Luyaba, as many as 80 percent of the inhabitants are involved in the commercial collection of native herb species, with this activity representing the primary source of income for many families. The greatest volume and variety of medicinal and aromatic plants are found in and extracted from the foothill region. Herbs are wild-collected over approximately 40 percent of the surface area of San Javier, primarily during the nine months of spring, summer, and autumn. A few species are even collected in the winter months, when the stems are bare of leaves; these include *Lippia turbinata* Griseb. (poleo), *Aloysia gratissima* (Gill. & Hook.) Tronc. (palo amarillo), and *Aloysia polystachya* (Griseb.) Mold. *(té del burro)*. All three of these species are important in the yerba mate compuesta industry.

An analysis of the aromatic and medicinal species handled by herb wholesalers in the town of La Paz (Dpto. San Javier) was prepared by Lagrotteria and colleagues in 1987 and communicated in a report to the Subsecretaría de Gestión Ambiental, Gobierno de Córdoba (Lagrotteria et al. 1987; Lagrotteria 1987). A total of 230 plant species were acquired by wholesalers. Dpto. San Javier yielded 131 of these, while 99 were obtained from other regions. Of the species that originated within San Javier, 32 were exotic species in cultivation

Table 10.1. Native medicinal and aromatic plants pressured by overcollection in Dpto. San Javier, Prov. Córdoba, Argentina.

Scientific Name	Common Name	Family
Achyrocline saturejoides (Lam.) DC.	marcela	Asteraceae
Adiantum raddianum Presl	culandrillo	Polypodiaceae
Aloysia gratissima (Gill. & Hook.) Tronc.	palo amarillo	Verbenaceae
Anemia tomentosa (Sav.) Swartz	doradilla	Schizaeaceae
Baccharis articulata (Lam.) Pearson	carquejilla	Asteraceae
Baccharis crispa Sprengel	carqueja	Asteraceae
Chenopodium ambrosioides L.	paico	Chenopodiaceae
Ephedra americana Humb. & Bonpl.	tramontana colorada	Ephedraceae
Ephedra triandra Tul. emend. J. H. Hunziker	pico de loro	Ephedraceae
Equisetum giganteum L.	cola de caballo	Equisetaceae
Hedeoma multiflora Benth.	tomillo de las sierras	Lamiaceae
Hypericum connatum Lam.	cabotoril	Hypericaceae
Linum scoparium Griseb.	canchalagua	Linaceae
Lippia turbinata Griseb.	poleo	Verbenaceae
Lycopodium saururus Lam.	cola de quirquincho	Lycopodiaceae
Margyricarpus pinnatus (Lam.) O. K.	yerba de la perdiz	Rosaceae
Minthostachys verticillata (Griseb.) Epling	peperina	Lamiaceae
Passiflora coerulea L.	pasionaria	Passifloraceae
Plantago spp.	llanten	Plantaginaceae
Usnea spp.	barba de piedra	(a lichen)

and 99 were naturally occurring native taxa. Of the 99 native species collected from the wild, 20 are today considered to be under serious pressure in the region as a result of uncontrolled collecting practices (table 10.1).

The harvest of native species is carried out by small, low-income family groups who sell the plants to local wholesalers, or *acopiadoras*. Knowledge concerning the location and collection of plants is passed down through generations by family members. In this sense, such knowledge truly represents the group's ethnoecology as it relates to an aspect of the commons that greatly affects their lives and livelihoods. With little sense of ownership, and subject to paltry compensation, the family groups employ collecting techniques that are highly extractive and therefore destructive to the ecology of the commons. Collectors often rip shrubs from the ground, roots and all, rather than selectively pruning the above-ground shoots. Fires, which burned 224,910 hectares in the province of Córdoba in 1996, are sometimes set to encourage the growth of desirable species. Based on interviews with collectors and acopiadoras, we estimate that each collector gathers approximately thirty kilograms of fresh herbs a day, working an average of four days per week. Our estimate for the total annual harvest of native aromatic and medicinal herbs in San Javier is 1,440 metric tons (calculated as dried herbs).

Relative to the value of herbs in subsequent steps of the market chain, the plant collectors receive relatively little compensation for the plants they deliver to the acopiadora. They are sometimes paid with food or scrip—that is, vouchers that can be used to make purchases in acopiadora-operated stores. When collectors attempt to use scrip to make purchases, store owners may insist on discounting its face value, further reducing their purchasing power.

Little postharvest processing of the herbs takes place in the countryside. Large quantities of dried herbs are shipped in bulk to cities such as Córdoba and Buenos Aires for final processing and packaging. Some local acopiadoras chop the dried herbs and package them in plastic bags suitable for retail sale, but for the most part little added value activity takes place in the rural sectors where the herbs are harvested.

Problems with the Current System of Harvesting and Marketing Medicinal and Aromatic Herbs

Threats to Native Species and Environmental Degradation

Natural populations of the commercially harvested plant species in the Sierras de Córdoba region are declining in geographical distribution and abundance. This is a result of both indiscriminate collection and widespread habitat destruction. The arrival of railroads in the region during the 1930s resulted in large-scale deforestation. Valleys of the Sierras once possessed extensive hardwood forests of *Prosopis, Schinopsis*, and *Aspidosperma* species that today persist only in small, relictual stands. Additional human activities that contribute to habitat destruction include use of fertile valley soils for the manufacture of bricks (the second most important economic activity in the region after wildcrafting of herbs); mineral exploitation (primarily marble, granite, and quartz); and burning of pastures and forests around homesteads and along roadsides to control populations of weeds and insects. Competition from exotic species also threatens some native populations of medicinal and aromatic plants. Many of these thrive as weedy species in the disturbed landscape.

Species of *Crataegus* and *Pyracantha* compete with *Lippia turbinata* and *Aloysia gratissima. Rubus* spp. overgrow herbs such as *Achyrocline satureioides* (Lam.) DC., *Margyricarpus pinnatus* (Lam.) Kuntze, and ferns such as *Adiantum raddianum* Presl. and *Anemia tomentosa* (Sav.) Swartz.

Removal of the forest cover and burning of vegetation have hastened erosion of soils by wind and summer rains. Within populations of native herbs and shrubs, erosion is also exacerbated by the common collecting practice of ripping entire plants up by the roots. Many of these species grow on the slopes of mountains and hillsides, and recently formed erosional gullies are a frequent sight in such habitats.

Systematic monitoring is needed to accurately quantify the impact of wildcrafting on native plant populations in the Sierras de Córdoba region, but one need only observe the stacks of dried herbs that overflow from trucks and fill the warehouses of the acopiadoras to appreciate the large quantity of plants being harvested each year. Some species are probably sufficiently viable to withstand continued wildcrafting, particularly if less destructive harvesting techniques are implemented. Others, however, have already become scarce and are threatened with local extirpation.

Economic Pressures and Lack of Quality Control

The low compensation received by collectors, the decline in abundance of native herbs, and the lack of alternative forms of employment have contributed to the structure of the current system of harvesting and marketing herbs, one that offers few opportunities to families of collectors for improving their economic status and few incentives for regulating the harvest at sustainable rates. The unskilled, poorly paid wildcrafters have little negotiating leverage with the acopiadora who purchases or barters for their harvest.

A related concern, one of the charges that is often leveled against producers of phytomedicines worldwide, is the lack of quality control. Analyses of herbal medications, even those that have been relatively highly processed and elaborately packaged, often reveal great variation in chemical composition or adulteration with other species. The current system for harvesting and marketing native herbs in Córdoba is remarkably unregulated. Adoption of more stringent standards and more careful documentation could greatly improve the safety and reliability of both crude preparations of aromatic and medicinal herbs and their more refined products. Many of the plants that enter the commercial market in Córdoba are not included in the two national registries of aromatic and medicinal herbs: the Farmacopea Nacional Argentina (FNA) and the Código Alimentario Argentino (CAA). Species included in these registers are accompanied by monographs describing their accepted use, identification, preparation, and potential risks. In 1995 nine species were added to the CAA list to bring it into agreement with codes in other Mercosur countries. The additions included canchalagua, carqueja, incayuyo, yerba lucera, marcela, peperina, poleo, vira-vira, and zarzaparrilla.

A survey by Lagrotteria and colleagues in the city of Córdoba recorded 475 species available commercially, marketed under 360 common names. Frequently one common name is used to refer to several different species, which adds to considerable confusion along the market chain and contributes to adulteration of herbal products. Of the 475 species recorded, only 147 are officially registered in Argentina—72 in FNA, 51 in CAA, and 24 in both lists. This means that 69 percent of the species marketed are not officially registered for use, although they are marketed side by side with the approved species. Two species that have been officially prohibited for sale by the Ministerio de Salud are *Chenopodium ambrosioides* L. and *Heliotropium curassavicum* L. The essential oil of the former contains a high percentage of the toxic compound ascaridole; the latter species contains pyrrolizidine alkaloids that are considered carcinogenic.

There is ample evidence that numerous native species not included in the official registers are safe for human consumption, based on centuries of traditional use. Several institutions in Argentina are pressuring the government to include monographs of these species in the FNA and CAA. At the same time, quality controls are necessary to guarantee the accurate identification of herbs in the marketplace, as well as acceptable standards of sanitation during harvest, processing, and packaging.

An Alternative Framework to Overextraction

Cultivation of commercially important plant species can provide an effective means of protecting natural populations from overcollection and local extinction. It can also create new economic opportunities for rural communities and improve the quality and reliability of products. PRODEMA (Programa para el Desarrollo de Plantas Medicinales y Aromáticas, or Program for the Development of Medicinal and Aromatic Plants) was established as a collaborative effort to develop horticultural production techniques and a marketing system for the sustainable harvest of medicinal and aromatic herbs in the province of Córdoba. Lagrotteria was the coordinator from 1994 to 1995. The principal objectives of PRODEMA were:

1. to introduce native species of aromatic and medicinal plants into cultivation, converting them from threatened species to an agricultural base for the development of rural communities;
2. to develop educational programs for collectors and training programs for producers of cultivated medicinal and aromatic crops;
3. to collaborate in the creation of laws to control collection of plants from natural populations and to establish quality controls for all herbs destined for human consumption;
4. to diversify regional economies in the province of Córdoba;
5. to reduce emigration from rural areas by developing and supporting small producers of aromatic and medicinal plants and facilitating their access to markets.

The provincial government, through the Subsecretaría de Medio Ambiente (Undersecretary of the Environment) and collaborating faculty at the Universidad Nacional de Córdoba supported PRODEMA as a means of diversifying the local economy and therefore reversing the trend of emigration from the Sierras de Córdoba region. To achieve these goals, and to serve as an effective model for similar programs elsewhere, PRODEMA was committed to developing a solid foundation based on scientific research. Education also played a central role in the program, including technical training for producers, development of research skills for students, and environmental education for families in the province. Development and implementation of PRODEMA required cooperation between research scientists, educators, government agencies, private industry, and local communities.

Córdoba's provincial undersecretary of environment was the first institution to endorse the creation of PRODEMA and to provide political and infrastructural support. The latter included a forty-five-square-meter chemistry laboratory for distillation of essential oils. The government also approved development of PRODEMA as a self-financing organization, to be supported in part by fees charged in return for services to its clientele groups (such as training courses and field inventories of native vegetation). Two private foundations, Fundación Juan Minetti and Fundación Juan Stabio, also provided support in the form of scholarships for students working on the project; funds for research conducted by faculty at the Universidad Nacional de Córdoba (including members of the biology, chemistry, and agronomy faculties); logistical support for field studies, including sites in the countryside for test plots; funds for international travel and training programs; and agricultural equipment.

Faculty at the University of Georgia in the United States collaborated through joint research projects concerning ecological, horticultural, and ethnobiological components of this highly interdisciplinary program.

That was the status of PRODEMA in 1995, when the conference that led to this volume was held at the University of Georgia. In 1996 government support for PRODEMA was severely cut back when the provincial government changed hands and economic and social priorities shifted. In response to this political change another primary supporter of the program, Fundación Minetti, redirected its support to public hospitals and maternity clinics, which the new government had also ceased to subsidize for lack of funds.

Today, the organizer and coordinator of PRODEMA, Marta Lagrotteria, continues her work in a new and productive context. With the support of the Universidad Libre del Ambiente, under the administration of the Municipio de Córdoba Capital, she continues to implement the objectives of PRODEMA by offering a training program for the initiation and development of small businesses dedicated to the production and sale of medicinal and aromatic species. During 1996, 120 students enrolled in courses that Lagrotteria offered through the Universidad Libre. Two levels of specialization have been initiated. The first provides a general introduction to the subject of aromatic and medicinal herbs. More advanced students specialize in production techniques (for example, greenhouse management), crafts, or culinary skills. Enrollment in these courses is heterogeneous: 40 percent of the participants are unemployed and are interested in developing job skills; 40 percent are students or professionals; the remaining 20 percent are interested in investing in herb-related businesses.

Collaboration with the University of Georgia continues and is focused upon the scientific objectives articulated earlier during the development of PRODEMA. Research examines several issues at the interface of ecology, botany, horticulture, conservation, and ethnobiology. The successful domestication of a new crop species requires knowledge concerning its life history characteristics, environmental tolerances, nutritional requirements, and susceptibility to pests and diseases (Mathe 1988). In the case of aromatic and medicinal crops, the chemical composition of the essential oils or active principles that give the plants commercial value are also of primary interest. All of these characteristics are likely to show some degree of variation in natural populations as a result of genetic variability and phenotypic responses. An understanding of these patterns of natural variation and their causes can greatly facilitate the sampling of natural populations for propagation and the design of breeding programs once plants have been brought into cultivation.

One of the current scientific tasks is to carry out more detailed field ecological studies and inventories to quantify the impact of current collecting activities on natural populations and to identify those species most suitable for domestication. This evaluation needs to be made because some species currently wildcrafted are sufficiently abundant in nature so that domestication is not necessary. Improved collection techniques to reduce plant mortality and long-term monitoring of populations should suffice to protect them as a permanent resource of the commons. Other species, as a result of their rarity, fragility, high market demand, or potential economic value, are priority candidates for domestication. For these species, detailed ecological studies in the field concerning soil, water, and light requirements; mechanisms of vegetative and sexual reproduction; and resistance to insects and diseases will greatly facilitate their development as new crops.

From an ecological standpoint, the primary advantage of cultivation versus wildcrafting of aromatic and medicinal herbs is conservation of natural plant populations and their habitats. From the commercial, public health, and consumer standpoints, the advantages of cultivating these plants as new crops are numerous. Palevitch (1991) and Bonati (1991) have summarized the advantages of cultivation over collection from the wild for medicinal plants; their observations apply equally well to aromatic plants wildcrafted for other commercial purposes. Cultivating medicinal plants increases their availability and maintains a more stable supply. Cultivated plants can be processed and preserved under more controlled conditions, increasing their quality. Botanical identification of cultivated plants is much more likely to be accurate than identification of wildcrafted plants; batches of the latter often include mixed collections of other species, or consist of species that are only superficially similar to the intended product. Plants in cultivation can be bred for improved genetic characteristics, including the quality and concentration of essential oils or medicinally active compounds. They can be manipulated agronomically to increase production and improve quality, using mulches, integrated pest management, irrigation, composting, and other organic production techniques. Finally, postharvest handling of plant materials is usually much better with cultivated than with wildcrafted plants, improving their preservation and hygienic qualities.

The value of most aromatic and medicinal species is based on their chemistry, a characteristic that can vary significantly in response to genetic and environmental factors (Charles et al. 1990; Fluck 1955; Hornok 1983). The oils distilled from aromatic plants (often called essential oils) are in fact complex mixtures of many different compounds (see, for example, Alkire et al. 1994). The components of these mixtures are under both genetic and environmental control (Lincoln and Langenheim 1976, 1978). Different individuals within a single natural population can display significant variation in essential oil composition. Faculty and students at the Universidad Nacional de Córdoba are analyzing the chemistry of native herbs to determine the degree of variation in natural populations and to identify individual plants with desirable chemotypes. Breeding programs will be developed using plants that have been introduced into cultivation to improve the quality, consistency, and yield of aromatic oils (or nonaromatic compounds of medicinal value, depending on the species).

Protecting and Profiting from the Commons

As a sustainable development initiative, the aromatic and medicinal plant program underway in Córdoba must confront numerous questions and conflicts related to the concept of the commons as communally shared property and resources. These include limitation of individual access to natural resources that have previously not been regulated or protected; restructuring of social and economic relationships; and clarification of intellectual property rights. This proposed conversion from an extractive economy to one based largely on private or cooperative agricultural production will force many communities to reconsider the boundaries and ownership of their regional commons.

This approach—combining applied research, training and extension programs, and small business development—attempts to protect the commons while at the same time increasing economic opportunities through three action principles.

DOMESTICATION AND IMPROVED HARVESTING TECHNIQUES TO PREVENT DESTRUCTION OF A COMMONLY HELD RESOURCE—NATIVE PLANT BIODIVERSITY. There are no effective incentives in the current system of wildcrafting to protect the long-term survival of native plant populations that are of economic value. In a comparable discussion of wildlife management options for the Chaco region of Argentina, Bucher (1989) cites the "tragedy of the commons" as defined by Hardin (1968). Profitable foreign markets for native game animals of the Argentine Chaco exist in several industrialized countries. In the Chaco, Bucher argues, wild game animals are considered public property and there is little interest in ensuring their sustainability, since "*si no lo exploto yo, otra lo hará*" ("If I do not exploit it, someone else will").

The base of the market chain for Chaco wildlife resembles the system described above for herbs in Córdoba: Campesinos hunt on their own property or someone else's, then the harvested game is passed through intermediaries to distributors who handle the sale and exportation of the product. In the case of native medicinal and aromatic herbs, the market is primarily domestic, but the system encourages campesino gatherers to collect as many herbs as they can from wherever they can be found. The decline in the availability of wild plants resulting from this type of overextraction is one of the factors that have contributed to the consistent trend of emigration from Dpto. San Javier during the last three decades. Domestication of the most threatened commercial herb species, training programs for wildcrafters in less destructive harvesting techniques, and scientific monitoring of native plant populations are all strategies designed to reduce overexploitation of native species and increase the range of options available locally.

INCREASED PROFITABILITY AND EQUITABILITY FROM RESOURCES. The current extractive economy based on the wildcrafting of herbs offers few opportunities for herb collectors to improve their economic position within the system. The family groups that harvest the herbs typically deliver all their harvest to a single acopiadora. Once the herbs are out of the collectors' hands, these people have no control over the remainder of the market chain. Current efforts encourage development of small farms and microenterprises that would open subsequent steps in the market chain (where there is significant value added) to greater numbers of participants, and encourage exploration of new production techniques and market niches.

Courses of study currently offered through the Universidad Libre were described above. Under the former auspices of PRODEMA, several short courses were organized and presented to local communities. The first course, offered in 1994, resulted in the formation of a cooperative, Las Sierras Ltd. Participants in the cooperative included people of low and middle incomes, including some landless families. All accepted the PRODEMA guidelines concerning harvest, production, and conservation of native herbs, and expressed their interest in working together to support expanded development of herb production. The cooperative was legally established in 1995 and completed several bulk sales of herbs within the first month.

PRODEMA organized additional courses, with the assistance of local governments and chambers of commerce, to train and organize potential producers. One program encouraged landowners with more than fifty hectares of land to adopt economical and sustainable techniques for harvesting herbs from their own property. If the harvested plants include mature fruits, participants agreed to provide the seeds to PRODEMA. Collaborators in the Facultad de Ciencias Agropecuarias at the Universidad Nacional de Córdoba used the seed to experimentally analyze factors controlling germination of commercially important native herb species. In turn, the researchers provided producers with improved knowledge and technology for bringing native species into cultivation. Another program for smaller landowners offered training in greenhouse production of herbs and "manufacture" of earthworm compost, a valuable organic fertilizer.

Today, in addition to increasing the volume and quality of harvested crude herbs, producers are encouraged to explore refining and marketing techniques that will bring more value-added income to local communities where the herbs are harvested. Distillation of essential oils, attractive packaging of crude herbs, and production of herbal cosmetics are all means of increasing the value of herbal products and diversifying local economies.

BETTER ORGANIZED SYSTEMS OF PRODUCTION AND REGULATION TO FACILITATE PROTECTION OF INTELLECTUAL AND GENETIC PROPERTY RIGHTS. The native flora in the Sierras de Córdoba region is a rich source of genetic diversity. As study and management of this diversity intensify through chemical and ecological analyses of native plants and the development of new crops, the monetary value of this biodiversity will also increase. Programs are already underway to screen folk medicinal plants in Argentina for antimicrobial activity (Anesini and Perez 1993). The International Cooperative Biodiversity Program (ICBP, a joint effort involving the U.S. National Science Foundation, the U.S. Agency for International Development, and the National Institutes of Health) funded a project to investigate the arid lands of Argentina, Chile, and Mexico for rare plants that contain chemicals with potential for making pharmaceutical drugs, as well as environmentally safe herbicides and pesticides. The more actively that groups in Argentina participate in these activities and the more judiciously they manage native germplasm, the greater the likelihood that they will receive a fair return on the economic value of the country's vast biological resources.

Preserving and documenting the natural diversity of the commons are important first steps. Subsequent steps, in the case of sustainable harvesting of medicinal and aromatic herbs, must deal with such questions as these: How can local knowledge of small-scale family collectors be used to guide further development efforts? What adjustments need to be made as people shift from wildcrafting for subsistence to quality-controlled harvesting for commercial purposes? How are profits to be channeled to reach those who traditionally have been the least benefited by extraction from the commons? What effects will all of these changes have on the integrity and management of the commons? The answers to these questions will not be easy to formulate. Clearly, ethnoecological research into these aspects of the commons issue can make a difference.

References

Akerele, O. 1991. "Medicinal Plants: Policies and Priorities." In *The Conservation of Medicinal Plants*. O. Akerele, V. Heywood, and H. Synge, eds. Cambridge: Cambridge University Press.

Alkire, B. H., A. O. Tucker, and M. J. Maciarello. 1994. "Tipo, *Minthostachys mollis* (Lamiaceae): An Ecuadorian Mint. *Economic Botany* 48:60–64.

Anesini, C., and C. Perez. 1993. "Screening of Plants Used in Argentine Folk Medicine for Antimicrobial Activity." *Journal of Ethnopharmacology* 39:119–28.

Bonati, A. 1991. "Industry and the Conservation of Medicinal Plants." In *The Conservation of Medicinal Plants*. O. Akerele, V. Heywood, and H. Synge, eds. Cambridge: Cambridge University Press.

Bucher, E. H. 1989. "Conservación y desarrollo en el Neotrópico: En busqueda de alternativas." *Vida Sylvestre Neotropical* 2:3–6.

Charles, D. J., J. E. Simon, K. V. Wood, and P. Heinstein. 1990. "Germplasm Variation in Artemisinin Content of *Artemisia Annua* Using an Alternative Method of Artemisinin Analysis from Crude Plant Extracts." *Journal of Natural Products* 53:157–60.

Cunningham, A. B. 1991. "Development of a Conservation Policy on Commercially Exploited Medicinal Plants: A Case Study from Southern Africa." In *The Conservation of Medicinal Plants*. O. Akerele, V. Heywood, and H. Synge, eds. Cambridge: Cambridge University Press.

Fluck, H. 1955. "The Influence of Climate on the Active Principles in Medicinal Plants." *Journal of Pharmacy and Pharmacology* 7:361–83.

Given, D. R. 1994. *Principles and Practice of Plant Conservation*. Portland, Oreg.: Timber Press.

Hardin, G. 1968. "The Tragedy of the Commons." *Science* 16:1243–48.

Hornok, L. 1983. "Influence of Nutrition on the Yield and Content of Active Compounds in Some Essential Oil Plants." *Acta Horticulturae* 132:239–47.

Lagrotteria, M. 1987. "Análisis del manejo y comercialización de plantas medicinales y aromáticas en la provincia de Córdoba." Report, Subsecretaria de Gestión Ambiental, Gobierno de Córdoba, Argentina.

Lagrotteria, M., M. Difeo, A. Toya, and R. Montenegro. 1987. "Situación de plantas y otros vegetales medicinales y aromáticos en la provincia de Córdoba." Sociedad Argentina para la Investigación de Productos Aromáticos, Publicación 8: 111–25.

Lincoln, D. E., and J. H. Langenheim. 1976. "Geographic Patterns of Monoterpenoid Composition in *Satureja douglasii*." *Biochemical Systematics and Ecology* 4:237–48.

———. 1978. "Effect of Light and Temperature on Monoterpenoid Yield and Composition of *Satureja douglasii*." *Biochemical Systematics and Ecology* 6:21–32.

Mathe, A. 1988. "An Ecological Approach to Medicinal Plant Introduction: Herbs, Spices, and Medicinal Plants." *Recent Advances in Botany, Horticulture, and Pharmacology* 3:175–205.

Palevitch, D. 1991. "Agronomy Applied to Medicinal Plant Conservation." In *The Conservation of Medicinal Plants*. O. Akerele, V. Heywood, and H. Synge, eds. Cambridge: Cambridge University Press.

Riddle, J. M. 1992. *Contraception and Abortion from the Ancient World to the Renaissance*. Cambridge: Harvard University Press.

Xiao, P. 1991. "The Chinese Approach to Medicinal Plants—Their Utilization and Conservation." In *The Conservation of Medicinal Plants*. O. Akerele, V. Heywood, and H. Synge, eds. Cambridge: Cambridge University Press.

11

Managing the Maya Commons

The Value of Local Knowledge

SCOTT ATRAN

Globalization often carries the assumption that an "invisible hand," which maximizes the common good by attending only to individual gain, is a natural condition of human psychology and biology. The invisible hand of microeconomics becomes hardly discernible from the brute genetic processes of evolutionary competition that supposedly regulate relationships within and between nonhuman species. Culture and cognition seem to arise only as "prosthetic" devices for the imperfect realization of these natural laws. This reduces the scientific role of anthropology and psychology to documenting, explaining, and correcting deviations from this rather particular cosmic ideal of "freedom."

There are, however, thinkers like Hardin (1968) who do not doubt the natural authority of self-interest, but show how its logic results in an eventual breakdown of any finite resource base. As long as there is no resource limit—as in a growing, colonizing, or globalizing system—the group's resources need not decline, and may well increase, as individuals seek only gain. But if common-pool resources become limited, and demand outpaces technology's ability to renew or extend them, the short-term quest for gain will in the long run destroy the group's resource base. For if each person expects the others to take on board the most they can, then it is in everybody's rational self-interest to do so while the ship is still afloat, even if it means eventually sinking with it.

The response to Hardin's bleak assessment of human ability to successfully manage a resource-limited commons has centered on numerous case studies of the social institutions that manage apparently successful local commons regimes (Martin 1992; Ostrom 1990). Foremost is the objection that

Hardin's "model overlooks the role of institutions that provide for exclusion and regulation of use" (Berkes et al. 1989:93). Societies develop institutional arrangements for closed access to free-riders (cheaters, invaders), so that local appropriators do not face the risk that the benefits produced by their efforts go to those who did not share in those efforts.

To this, Hardin (1991) and others respond that their concern is with open access, especially in regard to perishable or pollutable global commons, such as the earth's forests, ranges, waters, and air. In open access systems there are few, if any, effective means of preventing overexploitation by communities or nations. Witness the increasing conflict over diminishing North Atlantic fishing stocks between the United States and Canada, France and Spain, and the European Union and the North American Free Trade Zone.

A study of knowledge structures opens the possibility that there are cognitive means somewhat independent of institutional aspects for managing the commons, while also helping us to fathom how and why such institutions are even conceivable. In fact, the knowledge structures associated with the most sustainable commons behavior may be only loosely bound to local institutions of closed access, kinship, commensal obligations, or formal communication channels—or even to demography. If so, then the debate on what makes or breaks a commons, and on what the limits are to upscaling the lessons of local commons, may need rethinking. This is a hopeful prospect, at least compared with the alternative of starting from scratch in the face of geometrically increasing rates of worldwide pollution, rapid resource depletion, and massive species extinction.

This chapter explores implications for commons management in the light of one representative ecology task, with the aid of ethnography and a social network analysis of forest information channels. The aim is to better reveal the structures and values of local knowledge that make common survival on a limited resource base possible. This work is based on ongoing research at several Mesoamerican sites that is supported by the National Science Foundation grant SBR 94-22567 to the author and his colleagues.

Mayaland: Different Actors on the Commons Stage

Various groups—local Maya, immigrant communities (both Maya and Ladino), and nongovernmental organizations (NGOs)—have converged on forests in lowland Mexico and Guatemala. Each appears to have distinct views of how the forest works, what actions would destroy it, how it should be used, how (and among whom) its resources should be shared, and which of its resources are most valuable. What follows is one representative set of experiments that aim to reveal the kind of knowledge that can lead to action that sustains or destroys the forest commons. It is based on a fundamental assumption of cognitive psychology: that to understand how people put to use what they know in what they do, one must first of all seek to understand what they know.

The ecological study concerns three towns in the Municipality of San José in the north-central part of Guatemala's Department of El Petén: the native Itzaj Maya town of San José, the immigrant Ladino town of La Nueva San José, and the immigrant Q'eqchi' Maya town of Corozal. San José has some fifteen hundred inhabitants. About half identify themselves as Itzaj, but less than a tenth speak or understand Itzaj (a language of the Lowland Mayan family that includes Yukatek, Mopan, and Lakantun). All Itzaj adults in the study were bilingual in Itzaj and Spanish, the last monolingual Itzaj speaker having died a decade ago.

San José was founded by the Spanish in 1708, one of a handful of "reductions" for concentrating remnants of the native Itzaj population (and fragments of related groups). Although the Itzaj who ruled the last independent Maya polity were reduced to corvée labor after their conquest in 1697, their forests continued to thrive. Since 1960, when Tikal and other *ejido* lands were first alienated from the Itzaj, half the forest cover of Petén (which includes thirty-six thousand square kilometers, about a third of Guatemala's territory) has been cleared. In 1990, the Guatemalan government declared most of the remaining forests in northern Petén part of a UN-sponsored "Maya Biosphere Reserve" (including most former Itzaj ejido lands). This involved a "debt-for-nature" swap organized by the U.S. Agency for International Development, and an initial USAID plan to invest over ten million dollars in biosphere-related projects by 1996.

The neighboring town of La Nueva San José was established in 1978 under jurisdiction of the then-Itzaj-controlled municipality of San José in accord with the military government's plan to open the underpopulated forest lands of Petén to colonization. The vast majority of some six hundred immigrants in La Nueva are Ladinos—that is, native Spanish-speakers of mixed European and Amerindian descent—although there are a few Cakchiquel Mayan speakers from southern Guatemala. The town of Corozal, also under jurisdiction of the San José Municipality, was settled at about the same time by

Q'eqchi' speakers, Highland Maya from the Department of Alta Vera Paz just south of Petén. Many of the four hundred or so Q'eqchi' understand Spanish, and some speak it, but few willingly choose to converse in it. Q'eqchi' is not mutually intelligible with Itzaj.

Methods for Exploring Knowledge Structures

We asked Itzaj and Ladinos "which kinds of plants and animals are most necessary for the forest to live?" We also asked for explicit judgments concerning dependencies between plants and dependencies between animals, such as whether or not animals help (+1), hurt (−1) or do nothing to (0) other animals (Atran and Medin 1997). We followed through such studies with analyses of response justifications and both structured and open-ended interviews covering a whole range of topics. Notice, however, that even the most deceptively simple tasks and codings could yield rich information about relationships along several ecological dimensions: mutualist (+1,+1), commensalist (+1,0), parasitic (+1,−1), mutually destructive (−1,−1) and neutral (0,0).

Among the thousands of species scientifically present and the hundreds locally known, informants concentrated on a relatively small sample of approximately two dozen species of plants and two dozen species of animals. For plants these were overwhelmingly canopy trees, fruit trees, understory palms, and vines. For animals, these were largely predators and game. Nearly all the nominated plants and animals have high cultural use value; however, the reasons informants gave for their choices almost always involve ecological value. For example, xate (*Chamaedorea* sp.) refers to three generically related species of small palm that have no traditional significance but are nowadays collected for export; yet Itzaj informants stress that xate protects the forest floor, allowing other plants to thrive. Thus, social or economic value appears to render salient the ecological value of certain species, and these species are represented in both cultural and ecological terms. A salient group difference arises from the representation of interspecies relationships. Even for the few nominated integral species, like the jaguar in relation to other animals and vines in relation to other plants, Itzaj often see these relationships as reciprocal, whereas Ladinos view them unidirectionally.

We also collected social network judgments in all three groups. One reason for doing so is that we wanted to look at whether and how ecological knowledge might be socially transmitted. The Itzaj and Ladinos live in villages separated by just a few hundred meters, whereas the Q'eqchi' live more than an hour's distance by foot (Q'eqchi' agricultural plots, or milpas, however, are adjacent to those of the Itzaj and Ladinos).

The rationale for this experiment is that for the forest to survive, it is necessary on balance that interacting species not create catastrophic harm, or even help other species to regenerate. The dominant interactive species in the area for more than two millennia has been *Homo sapiens*. We anticipated that consensus on how people affect plants might well summarize the current state of actual ecological and economic relationships because this is the domain where the connection between knowledge and action is likely to be most salient and immediate. In addition to any overall differences, we were interested in focusing on the ecologically (and economically) dominant species.

Informants were six men and six women from each of three towns belonging to the Municipality of San José, Petén. For each population, instructions and responses were given in that population's native language: Itzaj, Spanish, or Q'eqchi'. To ensure the social diversity of each sample, no persons in the sample could have immediate kinship or marriage links with one another, where such immediate links are recognized as socially privileged by each group (parent/child, grandparent/grandchild, sibling, first cousin, sibling-in-law, parent/child-in-law, and god-parenthood, or *compadrazgo*).

To select plant stimuli, we asked this question: "Which plants and animals are most necessary for the forest to live?" We compiled a plant list of the twenty-six generic species (folkgenera) most frequently mentioned by informants. These were all trees, vines, or palms. To this list we added two often-mentioned life forms: grasses and herbs/underbrush. A blurred photograph of three people in local clothing with unidentifiable faces was used for the human item.

In this task, neither pictures of plants nor herbarium vouchers from our collections were used. Although at least half of the informants in each group were functionally illiterate, there were no differences in informant ability to reliably use name cards as mnemonic icons. This is consistent with previous work (Lopez et al. 1997). Vernacular and scientific names for the list of plants appear in table 11.1.

We asked each informant whether the people in their community help (+1), hurt (−1) or do not significantly affect (0) each item in the plant list. For each plant, we asked informants to justify their responses. We expected that this relatively simple task would reveal a detailed consensus on how people conceive of interactions with what they consider to be the most important forest plants. Note that the form of the question asks for an assessment of actual practice rather than ideal behavior.

In these methods, information from a number of informants is elicited. Rather than assume an underlying model for all informants from a given group, we test whether consensus exists to justify generalizing to a "culture." We adapt factor-analytic techniques from the "Cultural Consensus Model" of Romney et al. (1986). The model assumes that widely shared information reflects high concordance, or "cultural consensus," among individuals. To the extent that some individuals agree more often with the consensus on a set of related questions, they are more "culturally competent" than others.

Factor Analysis

No reliable overall Petén consensus emerged. Itzaj and Q'eqchi' showed clear consensus for within-group consensus when considered individually; Ladinos as a group showed marginal consensus: Itzaj (ratio of first to second eignevalue = 5.7:1, average competence = .79, variance explained = 67 percent), Ladinos (2.9:1, .72, 53 percent), Q'eqchi' (13.4:1, .91, 83 percent). Finally, Itzaj and Ladinos considered together showed consensus—that is, they share aspects of a mental model not shared by Q'eqchi'. A striking result is the departure of the Q'eqchi' from the other groups with respect to overall levels of interaction. The Q'eqchi' report no effect of humans on plants roughly 85 percent of the time. This compares with levels of around 10 percent in the other two groups. For half of the twenty-eight plants, none of the twelve Q'eqchi' informants mentioned any interaction with humans. This compares with zero for Itzaj and Ladinos. In short, on average the Q'eqchi' see very little effect of humans on plants. (The Q'eqchi' results take on special significance when compared with results of another study exploring the effects of plants on humans: Many of the plants that the Q'eqchi' report having no effect on are plants that at least eleven of the twelve informants gave uses for.)

Given that each population attained consensus in their overall view of how people affect plants, we can justifiably consider the mean score for each plant as that population's assessment of its overall impact on that plant. The mean score, or impact signature, for each plant ranges over an interval scale from entirely beneficial (+1.00), through neutral (0.00), to entirely harmful (−1.00).

There are large overall differences in impact signatures. Overall means were .40, −.01, and −.09 for the Itzaj, Ladino, and Q'eqchi', respectively. We

Table 11.1. Rankings of environmental impact signatures.

IMMIGRANT Q'EQCHI' MAYA			IMMIGRANT LADINOS			NATIVE ITZAJ MAYA			
Rank	Signature	Common Name	Rank	Signature	Common Name	Rank	Signature	Common Name	Scientific Name (In Itzaj Rank Order)
1	0.33	quano	1	0.75	ceiba	1	1.00	ramon	*Brosimum alicastrum*
2	0.25	corozo	2	0.58	pacaya	2	1.00	chicle	*Manilkara achras*
3	0.08	grasses	3	0.58	xate	3	0.83	cedar	*Cedrela mexicana*
4	0.00	amapola	4	0.55	allspice	4	0.83	ciricote	*Cordia dodecandra*
5	0.00	cordage vine	5	0.50	ciricote	5	0.83	mahogany	*Swietania macrophylla*
6	0.00	chapay	6	0.42	chicle	6	0.75	xate	*Chamaedorea* sp.
7	0.00	ciricote	7	0.33	madrial	7	0.67	ceiba	*Ceiba pentandra*
8	0.00	broom palm	8	0.33	ramon	8	0.67	guano	*Sabal mauritiiformis*
9	0.00	jabin	9	0.17	cedar	9	0.67	madrial	*Gilricidia sepium*
10	0.00	kanlol	10	0.17	guano	10	0.67	allspice	*Pimenta dioica*
11	0.00	madrial	11	0.17	grasses	11	0.58	amapola	*Pseudobombax ellipticum*
12	0.00	pacaya	12	0.08	mahogany	12	0.58	chapay	*Astrocaryum mexicanum*
13	0.00	allspice	13	0.00	amapola	13	0.58	corozo	*Orbignya cohune/Scheelea* lun.

14	0.00	pukte	14	0.00	water vine	14	0.58	broom palm	*Crysophilia staurocata*
15	0.00	ramon	15	0.00	corozo	15	0.58	pacaya	*Chamaedorea tepejilote*
16	0.00	santamaria	16	0.00	yaxnik	16	0.50	grasses	*Cyperaceae/Poaceae*
17	0.00	yaxnik	17	−0.13	pukte	17	0.42	chaltekok	*Caesalpinia velutina*
18	0.00	herb/underbrush	18	−0.14	chaltekok	18	0.42	jabin	*Piscidia piscipula*
19	−0.08	strangler fig	19	−0.18	santamaria	19	0.42	manchich	*Lonchocarpus castilloi*
20	−0.08	water vine	20	−0.25	cordage vine	20	0.25	santamaria	*Calophyllum brasilense*
21	−0.08	chaltekok	21	−0.25	herb/underbrush	21	0.17	herb/underbrush	(various families)
22	−0.08	killer vines	22	−0.33	broom palm	22	0.08	strangler fig	*Ficus involuta*
23	−0.25	ceiba	23	−0.44	jabin	23	0.08	yaxnik	*Vitex gaumeri*
24	−0.25	manchich	24	−0.50	chapay	24	−0.25	pukte	*Bucida buceras*
25	−0.25	xate	25	−0.60	manchich	25	−0.33	water vine	*Vitis tillifolia*
26	−0.58	chicle	26	−0.67	strangler fig	26	−0.33	cordage vine	*Cnestidium rufescens*
27	−0.67	cedar	27	−0.67	killer vines	27	−0.58	kanlol	*Senna racemosa*
28	−0.75	mahogany	28	−0.75	kanlol	28	−0.58	killer vines	(various epiphytes)

Note: For impact signatures, +1.00 to +0.33 = beneficial human impact; −0.33 to −1.00 = costly human impact.

performed an analysis of variance on the impact signatures for all plants for each population: $F(4,135) = 7.801$, $p = .0001$. This reveals that only the Itzaj differ significantly (Fisher PLSD, $f < .05$) from the other groups in considering people on the average much more helpful than harmful to plants: Itzaj (M = .40, SD = .45), Ladinos (M = −.01, SD = .42), Q'eqchi' (M = −.09, SD = .24).

First we ranked signatures for each group, then we divided the rankings into three levels: beneficial human impact (0.33 to 1.0), variable impact (0.32 to −0.32), costly impact (−0.33 to −1.0). Table 11.1 summarizes these results. Percentages in each category (beneficial, variable, costly) for each population are as follows: Itzaj (68, 18, 14 percent), Ladinos (29, 46, 25 percent), Q'eqchi' (4, 86, 11 percent). Only for Itzaj does the first category incorporate the majority of important forest plants. For Q'eqchi', the first category contains a single plant species.

Our purpose in comparing signature rankings was to see whether and to what extent people's assessment of their impact on forest plants reflects salient ecological and economic relationships. In order to interpret the content of the results, it is first necessary to briefly describe the prominent sorts of ecological and economic relationships that are readily and independently apparent in the area.

In this century, biologists have described at least seven prominent ecological associations for central and northern Petén (extending into the southern Yucatan peninsula and western Belize). The current terminology for these associations in the literature was established by the Maya projects of the Carnegie Institution (Lundell 1937). This technical terminology was borrowed from the local population. The name chosen for each association comes from the local Spanish vernacular, with the local Spanish name being itself a translation of the local Native Maya term. The term for each association below (Itzaj names are in parentheses) reflects the dominant species:

1. Ramonal *(u-k̰'aax-il oox)*, after the ramon or breadnut tree, *Brosimum alicastrum*. The most frequently encountered tree of the upland forest canopy (height = 30–40 m × diameter = 90–100 cm), it is called in Itzaj "the milpa of the animals" *(u-k̰ol-il b'a'al~che')* because of the many animal species that live on its fruit and leaves (Atran 1993a). It supports an inordinate quantity of orchids and epiphytes, bee hives, and ant and bird nests. Lowland Maya historically survived on ramon when maize was not available.

2. Zapotal *(u-k'aax-il ya')*, after the chicozapote tree, *Manilkra achras* (25–40 m × 25–140 cm). The second most frequently encountered tree of the upland canopy, it is characterized by a highly diverse plant association that grows on well-drained calcareous areas. Its fruit nourishes people as well as many birds and mammals. As with ramon, a long history of local use and selection may be in part responsible for the tree's current widespread distribution.
3. Caobal *(u-k'aax-il chäkäl~te')*, after the mahogany tree of the upland forest canopy, *Swietania macrophylla* (30–45 m × 1–2 m) For two centuries, this has been one of the most sought after sources of fine wood in the world. Occasionally found in dense clusters, an increasingly rare presence often characterizes mature high forest with minimal understory growth. Native Maya use the bark for curing leather and for treating foot fungus.
4. Cedral *(u-k'aax-il k'u~che')*, after the tropical cedar of the upland forest canopy, *Cedrela mexicana* (20–25 m × 100–150 cm). It is closely related to mahogany, but occurs more often in stands at higher elevations. Its branches have many orchids and epiphytes as well as nesting oropendolas. The dry branches and mature trunk (often partially hollowed by rot) harbor several species of woodpeckers and ants, termites, and even ocelots and jaguars. While its polished wood is less lustrous than mahogany, Lowland Maya call it "God's tree" *(k'u~che')* and prefer its redolent wood, which is easier to work into canoes, houses, masks, and altars.
5. Botanal *(u-k'aax-il xa'an)*, after the guano or cabbage palm, *Sabal mauritiiformis* (15–25 m × 20–25 cm). It is a prime source of shelter for people and animals. The immature palm is the most apparent plant in the understory, growing in areas inundated less than three months per year. Where mature, they form a distinct subcanopy association that includes emergent trees of both the upland and lower forest. The understory contains an abundance of grasses and broad-leafed plants. The flowers attract numerous varieties of bees and its panids of small black fruits are important to many birds and mammals. Its dead trunk remains erect for years, harboring woodpeckers, toucans, and parrots.
6. Escobal *(u-k'aax-il aj-kuum)*, after the broom palm, *Crysophila staurocata* (7–9 m × 8–10 cm). This solitary palm tends to dominate in wooded swamps of lower bajo lands that are not continuously inundated. Common upland trees are largely absent, except for stunted exemplars that can tolerate acidic and poorly aerated soils. Knotted lianas, suffrutescent vines, and epi-

phytes are omnipresent. Its panids of two to three hundred fruits feed many birds. Native Maya use the dried leaves for emergency thatch, and to make brooms, baskets and hats.

7. Corozal *(u-k'aax-il tutz)*, after the corozo palms, *Orbignya cohune* and *Scheelia lundelli* (readily distinguished only by the shape of their anthers). These are medium-sized palms (10–15 m × 150–200 cm) with solitary trunks, enormous leaves, and large panids (four hundred to one thousand fruits). Groves of these palms rapidly colonize areas disturbed by fire. These are often the only source of shelter for animals who pass through these disturbed patches of forest. The palm supports a large variety of orchids, insects, and snakes. Large rodents and peccaries feed on the fruit. Itzaj formerly boiled the kernels to produce oil used for cooking, to luster hair and body, and to fuel lamps.

The exploitation of three tree species for sale to international markets has largely driven this century's underdeveloped and extractive regional cash economy: chicle, mahogany, and tropical cedar. In the last decade, the three species of the xate plant *(Chamaedorea elegans, C. erumpens*, and *C. oblongata)* have also become important to Petén's cash economy. With the collapse of the natural chicle market in the 1970s, projects sponsored by NGOs to preserve remaining forest through sustainable exploitation of nontimber resources have increasingly focused on xate (Nations 1992). Unlike other cash-producing species, however, xate has no traditional economic role to speak of, nor does it dominate any ecological association. Although men in all local Petén populations provide xate plants to outside contractors, local people are unaware of where the plants go or what they are used for (mostly for floral arrangements in Florida). Most local folk believe that the plant's exchange value derives from a hidden power, discovered by Western experts, to help (aphrodisiac) or hinder (contraceptive) one's sex life.

The fruit of the allspice tree has in recent years also become a source of cash for those Ladino and Itzaj *pimenteros (aj-men nab'a'~k'u'uk)* who often also work as *chicleros* (*aj-men cha'*) and *xateros (aj-men ix-xyaat)* in the forest. For the Itzaj, the leaf and fruit also have important medicinal properties against intestinal illnesses and for aiding in childbirth. Finally, another important tree for household economies is ciricote, prized both for its fruit and hardwood.

Content Analysis

For the Itzaj the ten ecologically dominant and economically most important plants all fall within the first category—that is, human impact on them is considered beneficial. Ramon and chicle both have a signature that is entirely beneficial, followed by mahogany, cedar, and ciricote, xate, guano, allspice, ceiba, and madrial. The ceiba tree was traditionally the sacred Maya tree of life and is today protected by law as Guatemala's national tree. The madrial tree is a prime source of durable hardwood. Its leaves are edible boiled or fried and served in many of the same recipes as the pacaya, and its flower is used as a nosegay along with the flower of the allspice tree.

Itzaj rank the pacaya palm, the Waree cohune palm, and the amapola tree along with broom palm and corozo palm. Pacaya leaves are occasionally used as Palm Sunday branches, and the edible spadices of the pacaya and Waree cohune constitute a regular part of the diet. The amapola tree is an important source of wood and, like mahogany, it can be used for making canoes when there is no tropical cedar available. Itzaj also include in their first category grasses as well as the leguminous hardwood trees. Grasses are used for wall construction, fodder, and medicine, and also provide food and shelter for grazing animals and a privileged hunting ground for snakes, felines, and other predators. Itzaj consider the leguminous hardwood trees to be folk-botanical allies that occupy similar roles in both the economy of nature and the economy of humans.

For Itzaj, the second category consists of another hardwood tree, Santa Maria, herbaceous undergrowth, the strangler fig, and the marginally useful but ubiquitous forest trees yaxnik and pukte. Itzaj consider herbaceous undergrowth and the strangler fig to have both positive and negative qualities. The fruits and descending growth of the strangler fig nourish and shelter a host of animals, and its roots are an emergency water source for people, but the strangler also kills other trees. Likewise, herbaceous undergrowth provides shelter and nourishment to animals, as well as food and medicine to people; however, it must be cleared regularly for people and trees to thrive.

In the third category, Itzaj include vines that provide emergency water and cordage because to use them is to kill them and because they are too plentiful to need protection. Last in ranking are the kanlol tree and strangler vines. The yellow-flowered kanlol is a most prominent tree in Petén, but it is considered a "weed tree." Itzaj recognize that strangler vines (mostly epiphytes) help to distribute water to other plants, but they are a general nuisance to people and trees.

Ladinos give top rank to the ceiba, but also claim to protect ramon, chicle, and other plants highly valued by Itzaj. Ladinos also include in the second category, which is the largest, plants highly valued by Itzaj. Unlike the Itzaj, their lowest category includes the most frequently used leguminous hardwoods.

For Q'eqchi' only the guano palm falls in the first category. The only other plants that have a marginally positive signature are the corozo palm and the grasses. Guano and corozo are both used for thatch, and grasses are occasionally planted for fodder. Q'eqchi' occasionally collect corozo fruits to eat or to sell to a local NGO. The vast majority of plants fall into the second category, where human impact is neutral regarding them. Lowest ranked are the three most important sources of cash income in Petén: chicle, tropical cedar, and mahogany. The ceiba, a leguminous hardwood, and xate fare only marginally better.

Only the Itzaj seem to have a vision of the role of humans in helping the forest to survive that is decidedly positive overall as well as ecologically and economically coherent. For Ladinos, the role of humans is neither decidedly positive nor negative overall. Nevertheless, there is a tendency to see humans as playing a positive role in preserving the ecologically dominant and important plants. Finally, Q'eqchi' have a marginally negative overall appreciation of the role of humans in preserving crucial forest plants. This is markedly negative for the economically most important plants.

The Q'eqchi' are distinctive in seeing humans as one sixth as influential on plants as the other two populations. This difference does not arise from unfamiliarity—another experiment (not reported here) revealed Q'eqchi' use of nearly all the plants. There is an important implication: Q'eqchi' see themselves as neither helping nor hurting most plants, as if plants are not affected by human behavior. This is a striking observation, given the evidence that this group has the most negative effect on plants.

Because of the overlap between some plants that are both ecologically dominant and economically most important, the ecological consequences of the Q'eqchi' vision should be contrary to those implicit in the Itzaj vision. One might expect short-term degeneration of the forest rather than long-term regeneration. There is some independent confirmation of this. On the one hand, tracking by remote sensing of Q'eqchi' expansion shows rapid and extensive deforestation along the recent migration routes (Steven Sader for Conservation International, 1996). On the other hand, in the last few years Itzaj have set up the only self-managed forest reserve in Petén (Atran 1993b), where there is already evidence of regeneration of plant and animal stocks in areas recently depleted by logging operations, by immigrant slash-and-burn agriculture, and by depredations from transient hunters.

The Itzaj and Q'eqchi' are fairly clear and consistent, leading to an expectation that Itzaj are likely to act to sustain the forest and that Q'eqchi' are likely to act in ways that will degrade the forest. The expectation for Ladinos is less clear. Before turning to a fuller discussion of the issue, it will help to compare social structures.

Communication Networks for Forest Information

The methodology of social network analysis can help us to understand various socially related structures (such as economic behavior) as well as social change (Granovetter 1979). More specifically for our purposes is the idea that "core" networks of close personal ties may interact with "extended" networks involving more distant ties to forest experts, so as to determine the community's overall ability to encounter and assimilate new information (Ford 1976; Hammer 1983).

For each community we began with twelve informants (six men and six women) over the age of thirty-five who were not immediately related by kinship or marriage. Each informant was asked to name, in order of priority, the seven people outside of the household "most important for your life." They were asked in what ways the people named in this social network were important for their lives. Later, each informant was also asked to name, in order of priority, the seven people "to whom you would go if there were something that you do not understand about the forest and want to find out about." They were asked about the kind of information they would seek in their expert network.

The decision to restrict the number of people named in each case to seven was based on previous cross-cultural studies. These indicate that personal networks of four to eight intimate relationships are readily elicited, but network structure can vary significantly across cultures and across different populations within a culture (Wellman 1979). A pilot study was undertaken to ensure that informants had no problem with the tasks. These were carried out in the language of each population. For each informant, we recorded on a coding sheet age, gender, occupation, relation to subject, ethnicity, community of residence, and frequency of contact with the person named.

After performing the tasks with our initial informants, we used a "snowball method" to extend these ego-centered networks to the wider context of patterned social communication in which they operate. Social interaction and expert networks were elicited from the first and last persons named in the social network (whose mother tongue might differ from that of the original informant). On the few occasions where either the first or last person named was not available, we interviewed either the second or sixth person named. The decision to establish network closure in this way was based on feasibility and on previous studies suggesting that in practice it is rarely necessary to inquire into direct ties involving more than one intermediary (Mitchell 1969).

The average age of persons named in each group was forty years or more, with about twice as many males named as females in each population. The one-step snowball method allowed for thirty-six interviews per population (informant + first named + last named x twelve), provided that no names were repeated across informants. The number actually interviewed proved about the same for Itzaj (30) and Ladinos (29), but much less for Q'eqchi' (20). The total number of individuals named by the Q'eqchi' (58) was also less than half the number named by Ladinos (110) or Itzaj (136). Thus, the overall network density (*Dh* = ratio of possible names to actual names) was far greater for the Q'eqchi' (4.55) than for Ladinos (2.4) or Itzaj (1.94). This was also true for measures of centrality and interconnectedness (Lambda Sets).

The high density and interconnectedness of the Q'eqchi' social network favor diffusion of information through multiple pathways and rapid assimilation by the whole of the community; however, this network's high centrality also suggests the focusing or screening of information by a few central actors. The Q'eqchi' expert network appears to be radically disconnected from the social network. Thus, whereas the social network consists wholly of Q'eqchi', the expert network is dominated by outside NGO and government institutions.

The greatest overlap between the social and expert networks occurs with Itzaj and the least with Q'eqchi'. For the Q'eqchi', only six of the most cited social partners (chosen three or more times) are among the eighteen most cited forest experts, and they are cited much less often as experts than outside institutions. The two highest ranked experts (each named by 60 percent of those interviewed) are a Washington-based NGO and the government agency managing the Maya Biosphere. For Itzaj, fourteen of the most cited social partners are among the twenty-two most cited forest experts. Although the Itzaj social network is not highly centralized, the most cited social partner is also the second most cited forest expert, and the top forest expert is also the third most cited social partner. Over two-thirds of the Itzaj agree that these two are the top experts.

For the Itzaj, there is a much less centralized and densely connected social network, with information liable to be diffused and assimilated in less concentrated or redundant ways. Still, any expert information pertinent to the forest is more likely to be diffused and assimilated through society. This is because the social and expert networks overlap significantly, and the top forest experts also tend to be key nodes in the social network—gateways along intersecting paths of information.

The Ladino social network resembles more the Itzaj than the Q'eqchi'. It is a bit more centralized and interconnected than the Itzaj, with church membership having a higher profile as grounds for interaction than among Itzaj. As with the Itzaj, forest experts from the community also figure prominently in the social network. For Ladinos, eleven of the most cited social partners are among the twenty-five most cited forest experts. Two of the three experts named the most (by more than 40 percent of those interviewed) are the same two persons that the Itzaj themselves most name as experts, and the third expert Ladino is father-in-law to the top Itzaj expert's daughter. Moreover, Ladino experts, who tend to mention Itzaj as their experts, are those persons most socially interconnected (the highest Lambda-Set level) in the Ladino community.

The Q'eqchi' networks suggest that information pertinent to the long-run survival of the forest comes from outside organizations that have little long-term experience in Petén. Moreover, what outside information there is does not seem likely to penetrate deeply into the Q'eqchi' community, nor is it likely to be perceived as relevant to their (social) lives. The Q'eqchi' do not know who the experts are in a socially relevant sense. For the Itzaj, by contrast, expert information about the forest appears to be integrally bound to the more intimate patterns of social life, as well as to an experiential history traceable over many generations, if not millennia.

For the Ladinos, expert information is also likely to be assimilated into the community. Because Ladino experts are also socially well connected, information that may come through Itzaj experts has access to the greatest number of interaction pathways. Ladinos know who the experts are among them; but these experts, and much of the rest of the Ladino population, also know that ultimate expertise lies with the Itzaj. In sum, Ladinos have the social connections to Itzaj expertise that may enable them to assimilate as much Itzaj knowledge as is compatible with their own experience and understanding.

Latent Knowledge Structures and the Maya Landscape

In the Itzaj, we have a society with no apparent tradition of institutions established to close off or monitor access to the commons, and without benefit of a higher authority to compel preservation of the environment for all. Yet Itzaj plainly think and act in ways geared to sustain their environment in the long run. They do more to preserve the environment than other groups who share it, although nearby communities have lesser populations but comparable access to common-pool resources. Itzaj seem to do this with less reliance on kinship and densely connected social networks than do the Q'eqchi'.

Historically, Lowland Maya readily could, and often in hard times did, leave family, friends, and village for different ones. New social attachments were relatively easy to establish (often by *compadrazgo*, or god-parenthood). There were also rendings of the community fabric owing to political repression, famine, or epidemics. Older Itzaj, for example, remember relatives having to flee, and some even to settle, in Soccotz on the other side of the border with Belize, to escape the harsh corvée labor imposed during the dictatorial reign of General Jorge Ubico in the 1930s (see Schwartz 1990 for other examples).

This fluidity of association and the proven ability of a Maya nuclear family to survive on its own in the forest for seasons (and if need be for years) would seem to encourage "individualistic" rather than cooperative behavior. Indeed, when we ask members of our three populations "How did you learn about the forest?" as often as not Itzaj (but not Ladinos or Q'eqchi') respond: "I walk alone" (*ḳ-in-xi'mal t-in-jun-al*, where "walk" *[xi'mal]* also means "to observe and behave appropriately").

Nowadays, Itzaj rarely coordinate hunting or planting among themselves, although in the recent past they collectively performed rituals for these activities. Not until 1991 was there any institutional setup for monitoring and closing access to the forest commons. Only then was a committee of Itzaj formed to establish the Bio-Itzaj reserve in order to preserve the last large tree stands in the municipality from loggers who had marked them all for cutting and from immigrants who had begun to burn fields on either side of the approaches that the loggers had made into the area. Closing access also created internal divisions.

This raises another puzzle. If in fact native Lowland Maya are self-sufficient, then, how is it that they have been so apparently successful in collectively sustaining both their society and environment over the centuries? And why do they believe that their fate is collectively bound to the forest? As one keen student of Lowland Maya history notes, "It is, however, no less a paradox than the emphasis placed on the autonomy and efficacy of the individual in Western urban industrial societies, where all but a few eccentrics who subject themselves to 'wilderness training' would quickly perish if deprived of the goods and services furnished by their fellows" (Farriss 1984:132). The paradox is genuine, but the answers proposed are not complete, whether in terms of "emotional comfort" or "material benefits" gained in "sustained interaction with one's fellows" (ibid.). It is not so much material sharing in a society which, apart from marriages and other festivals, nowadays lacks routine villagewide commensal institutions. Rather, it is the easy sharing and feedback of knowledge that allows an individual an effective, context-sensitive evaluation of resources and responses in an environment in flux.

To begin to account for Itzaj environmental awareness and behavior toward the forest, we will explore a kind of belief system that may be deemed a latent knowledge structure. Latent knowledge structures are cognitive procedures that make possible an evaluation of the future consequences of purposive behavior for one's life and for the environment in which life is embedded. A latent knowledge structure resembles a theory in its ability to take particular experiences and give them general relevance—that is, they take an instance of experience and project it to a (perhaps indefinitely) larger ensemble of complexly related cases (Wisniewski and Medin 1994). Unlike theories, however, such belief systems do not determinately specify relationships between entities, nor do latent structures directly imply or logically entail necessary or probable consequences of actions. Rather, such structures implicate a wide-ranging network of relationships between entities that anticipate a variable and somewhat open-ended range of responses and future consequences of actions. There are no stipulated, conventional, or systematic formulations of laws, standards, or methods. The function and significance of such beliefs is not ever-growing clarity, prediction, or knowledge advancement, but fitting the forms of human life to the ecological surroundings with which they are involved.

In the Itzaj case, there is no principle of reciprocity applied to forest entities, no rules for appropriate conduct in the forest, and no controlled experimental determinations of the fitness of ecological relationships. Yet, reciprocity is all-pervasive and fitness enduring. The latent knowledge structure of the Itzaj is robustly coherent in its implications and effective in its consequences, much as in the weaving and knotting of an Itzaj hammock from the hair-thin fibers of the henequen plant *(Agave fourcroydes)*: "As in spinning a thread we twist the fiber on fiber. And the strength of the thread does not reside in the fact that some fiber runs through its whole length, but in the overlapping of many fibers" (Wittgenstein 1958).

We are only beginning to discern some (1) basic elements (for example, generic species), (2) relational components (such as human, plant, animal interactions) and (3) network contours (for instance, the spirited cultural landscape of human/animal/plant intercourse) of these knowledge structures. It is this latter aspect that appears to be most crucial and most elusive for understanding the sustainability of "Mayaland." Places in the cultural landscape that Itzaj call "Mayaland" *(u-lu'um-il maayaj)* tag episodes in a person's life that Itzaj are most readily willing to communicate to others. Reference to such places "automatically" makes recounting of a personal experience culturally relevant for any other Itzaj who is listening—that is, individual and social identity are simultaneously implicated and constructed through such references. Itzaj place names also often describe ecological associations. Not only do these places bring together social history and autobiography but they are also privileged meeting grounds for humans, plants, animals, and spirits.

Nearly every place that Itzaj know and name in Petén is imbued with a sense of time/space distinct from a chronological sequence or a spatial map. Itzaj do not locate these places only, or even primarily, by spatial positioning any more than they describe their own lives in terms of temporally defined sequences of events. Rather, Itzaj refer episodes to these cultural loci that connect the different paths of individual life histories. Nor do Itzaj describe spirits or species in general terms, such as "All x is/does such-and-such." Instead, diverse episodes are interwoven into a more generalized cultural landscape where "everything has its place-and-responsibility *(ḵuch)* in the world" *(tulaḵal yan u-ḵuch yoḵ'ol ḵab')*.

We asked the individuals interviewed in our social network study whether they knew stories about the forest involving spirits and, if so, to recount an example. We also asked whether such stories were "true." Nearly every Itzaj recounted true stories about forest spirits, such as the one about the arux.

They recount that the arux play tricks on people to test the strength *(muḵ')* of a person's blood *(ḵ'iḵ'el)*, in order to assess valor in doing what is appropriate in the forest. The unworthy will show themselves too unsettled by such pranks and will start behaving badly in the forest: cursing at the wind, shooting too many animals, or cutting down too many trees. Then the arux will abandon them to their fate. But if they show respect, the arux will assist them, leading them to animals or to a stand of fruiting ramon trees or chicle trees full of sap. Such spirits do not have simply a "figurative" or "metaphorical" existence, but an affective presence that Itzaj will literally stake their lives on.

The Q'eqchi' knew virtually no stories, whereas Ladinos told stories they had heard from the Itzaj about the ability of Itzaj sorcerers to turn themselves into animals. The Itzaj call such sorcerers *aj-waay*. Although today Itzaj consider the waay to be demonic human souls who practice black magic, in pre-Columbian times the waay apparently referred to the particular animal "soul-mate" that made up part of each individual Maya's personal and social identity (Freidel et al. 1993).

The web of forest life in Mayaland is spirited with affective value that sustains reciprocity, respect, and fitness, and that goes beyond mere observation and consideration of the entities involved: "It is adaptive for cognized models to engender respect for what is unknown, unpredictable, and uncontrollable, as well as for them to codify empirical knowledge. It may be that the most appropriate cognized models, that is, those from which adaptive behavior follows, are not those that simply represent ecosystemic relations in objectively 'correct' material terms, but those that invest them with significance and value beyond themselves" (Rappaport 1979:100). The overlapping and criss-crossing network of purposive dependencies may help to account for the historical lack of institutional arrangements designed to close access to the forest commons. The Itzaj tendency to believe that when someone shows disrespect for the forest all are destined to suffer implicates a communicative network of animals-plants-people-spirits that is liable to make redundant direct human-human decisions to cooperate with one another in maintaining the common forest.

If the Itzaj lack of corporate ceremonies, institutions, and cooperative work based on kinship pose a problem for previous analyses of the commons, then the presence of these institutions among the Q'eqchi' poses an even more difficult problem. Of the twenty-three Mayan linguistic groups in Guatemala, the Q'eqchi' constitute the largest and most homogeneous, in terms of the smallest number of dialects and largest number of monolinguals (Stewart 1980). From 1960, when the military government opened Petén to colonization, the Q'eqchi' have been the largest identifiable group of immigrants, and the most isolated.

Unlike the Itzaj, the Q'eqchi' of Corozal practice milpa agriculture in kin groups, seeding together up to eight *manzanas* (about five and a half hectares) for each family of six persons. Q'eqchi' clear-cut and burn forest for milpa plots in order to grow three staples: maize, beans, and squash. New plots are contiguously burned every year (after a single crop) in a pattern that snakes through forest or secondary vegetation. Q'eqchi' do not keep forest reserves or protect trees and hilltops, nor do they indicate any need to do so when directly asked.

Itzaj also slash and burn to prepare milpa plots, but they do not cut valuable trees, which they surround with firebreaks when the rest of the patch is burned (Atran 1993a). They also ensure that trees continue to ring milpa plots and that hill crowns are neither cut nor burned, thus facilitating forest regeneration (see Remmers and Ucan Ek' 1996 for similar practices among Yukatek). In the area that any given native farmer exploits, the bulk of his parcel is usually left as a forest reserve (two to five parts forest to one part milpa). An average Itzaj family seldom clear-cuts land entirely, or uses more than five manzanas (about three and a half hectares) at a time for milpa. A man (almost never a woman) plants a multiplicity of crops. This results in milpas emulating and sustaining the biodiversity of the surrounding forest (see Nations and Nigh 1980 for Lakantun Maya farming). Land is fallowed less often and for longer periods than with Q'eqchi'. Plots tend to be smaller and, in some cases, the same plot is continuously cultivated for decades by mulching instead of burning.

Yet ceremonial life is manifestly richer among the Q'eqchi' than the Itzaj, including ceremonies related to agriculture. Indeed, when Itzaj invited members of another Q'eqchi' group to visit the Bio-Itza, the Q'eqchi' expressed surprise at the lack of Itzaj ceremonies and volunteered to teach the Itzaj some of their own (Atran 1993b). Nevertheless, the Q'eqchi' do not consider any elements or features of the Petén landscape to be sacred or to need protection, except for the *pom* tree *(Protium copal)*, whose sap provides the incense that mediates ritualized communication with the supernatural world (but which is not considered crucial to the survival of the Petén forest).

What is surprising is that Q'eqchi' do consider the cultural landscape from which they originated as sacred and its elements to have protected values. Many Q'eqchi' in Petén believe, like those who have remained in their homeland around Coban, in *tzuultaq'a*, or the sacred "mountain valley" (Wilson 1995). Such belief is especially current among elderly Catholics, just as beliefs concerning the spirited landscape of Petén are most strong among elderly Catholic Itzaj. All of the mountains around Coban are named, and different towns consider that they have ancestral ties to different mountains. Appropriating land requires sacrifices to the mountains, whose forests are protected.

In short, the affective involvement of the Q'eqchi' with the landscape of their homeland resembles Itzaj involvement with Petén; but little of it carries over from Coban to Petén. As one NGO operative reported when he tried to encourage the Q'eqchi' to stress the same concern for protection of nature that he had witnessed in Coban in order to better meet government criteria for gaining a concession in the Maya Biosphere: The Q'eqchi' responded that "in the mountains [of Coban] we use the land with God's permission, but not in Petén," so that their only interest was in gaining a concession wholly given to agriculture.

Unlike the Q'eqchi', but like the Itzaj, the Ladinos express a need to protect hilltops and valued trees, although less consistently and less amply than Itzaj. Like Q'eqchi', Ladino farmers cultivate on the average less than half as many kinds of crops as Itzaj. Their milpas tend to be smaller than those of the Q'eqchi' but larger than those of Itzaj. They practice no ceremonies that are directly linked to nature, as do Q'eqchi' (and, to a much lesser extent, the Itzaj); however, their church life helps to provide them an enduring sense of community that Q'eqchi' and Itzaj acquire by other means.

Unlike Q'eqchi', the Ladinos are learning aspects of Lowland Maya awareness, such as use and concern for the ramon tree, which is hardly known to Highland Maya and little appreciated by Q'eqchi' in Petén. The emerging Ladino crop patterns and tree valuations seem to reflect Itzaj attitudes and practice rather than attitudes and techniques proffered by NGOs or government extension workers. One wonders to what extent UN, NGO, and government activists even distinguish among "locals" (Arizpe et al. 1996).

In fact, there is a long history in Petén of Ladinos learning Itzaj agroforestry practices. In some Petén communities that are now entirely Ladino but were formerly mixed Ladino-Maya, even ancient Lowland Maya practices are readily discernible (Rice and Schwartz 1992). In fact, some of these longstanding Petenero Ladino communities perform variants of Maya rituals, such as the ceremony for rainstorms *(chaak)*, which Itzaj abandoned a generation ago. These Peteneros both express and practice reciprocity in dealings with the forest (Schwartz 1995). In brief, Mayanization of the Ladinos in Petén is part of a centuries-old process. As pervasive and profound as Hispanization of the Maya, it is, however, more subtle and less forced.

And here we come to our final puzzle: We have suggested that Itzaj are heedful of the forest by virtue of an affective and intentional cultural commitment to the Petén landscape, just as Q'eqchi' are heedless of the Petén forest because their culture is committed to another landscape. We have also suggested that the Ladinos are learning to take heed of the forest because of their social relationship with the Itzaj. Mere proximity and exposure to Itzaj practice would not seem sufficient for learning to attend to reciprocal relationships between humans and plants and between animals and plants, much less to Itzaj forest spirits. This intimates that the Ladinos are appropriating relevant aspects of Itzaj culture without any prior understanding of its language, history, or traditions. Ladinos, in fact, claim that they are learning from and about the Itzaj; yet the Itzaj disavow teaching the Ladinos anything about the forest. How, then, could Ladinos be learning Itzaj sensibilities? The tentative line of reasoning that follows should motivate further research.

Seeking to interview the two most cited Itzaj experts, we found that both had gone to the Ladino town of La Nueva. When they returned we asked them in separate interviews if they had ever taught anything about the forest to the Ladinos; both denied having done so. Then we asked why they had gone to La Nueva and what they did there. One said that he had gone because there were no lemons to be found in San José but he knew of some in La Nueva. He said that he had stayed in La Nueva after finding the lemons because he was trying to figure out with the people there where and how it would be best to plant lemon trees. The other Itzaj said that he had gone to visit his daughter, who is married to the son of the most cited Ladino expert. There he stayed behind, telling stories of the barn owl (*aj-xooch'* = Tyto alba), whose call augurs the death of strangers.

Ladinos are being educated by any variety of means—observation, stories, discussions—to what they ought to attend to. Because the Ladinos do not have all the instruments of Itzaj culture, and are unlikely ever to own them, there may always be doubt as to what meaning the forest may hold for those in the know. But the Itzaj, as a constant presence and reminder of such doubt, may be a powerful spur to attention.

Conclusion

The case of the immigrant Q'eqchi' Maya suggests that a vibrant indigenous language, a robust corporate ceremonial life, convivial cultural institutions, and cooperating kin are neither singly necessary nor collectively sufficient for a successfully managed small-scale community commons. Together, the native Itzaj Maya and Q'eqchi' cases indicate that an affective network that intentionally relates the elements and places in the commons landscape to a personalized cultural history can be paramount. The case of the Ladino immigrants, however, intimates that such motivating attachments to sustainability may be acquired despite a lack of predisposing cultural traits—and even perhaps because of such a lack.

What moral has this for upscaling human knowledge to face the daunting problems of more global commons? Possibly, with science and sensitivity, we may better understand how different people value the parts of the world they desire and keep for their well-being. Perhaps, inspired by doubt and, like the Ladinos, we may be educated by a process of inquiry and acquaintance to cumulatively heed aspects of nature and knowledge that we hitherto did not imagine. For what are the alternatives? To seek in an economic theory of substitutable values and insatiable gain our common ground, or to start from scratch? Surely, the hope is worth a try.

References

Arizpe, L., F. Paz, and M. Velázquez. 1996. *Culture and Global Change: Social Perceptions of Deforestation in the Lacandona Rainforest in Mexico*. Ann Arbor: University of Michigan Press.

Atran, S. 1993a. "Itza Maya Tropical Agro-forestry." *Current Anthropology* 34:633–700.

———. 1993b. "The Bio-Itza." *Anthropology Newsletter* 34(7):37.

Atran, S., and Medin, D. 1997. "Knowledge and Action: Cultural Models of Nature and Resource Management in Mesoamerica." In *Environment, Ethics, and Behavior*. M. Bazerman, D. Messick, A. Tinbrunsel, and K. Wayde-Benzoni, eds. San Francisco: New Lexington Press.

Berkes, F., D. Feeny, B. McCay, and J. Acheson. 1989. "The Benefit of the Commons." *Nature* 340:91–93.

Farriss, N. 1984. "*Maya Society under Colonial Rule.*" Princeton: Princeton University Press.

Ford, R. 1976. "Communication Networks and Information Hierarchies in Native American Folk Medicine." In *American Folk Medicine*. W. Hand, ed. Berkeley: University of California Press.

Freidel, D., L. Schele, and J. Parker. 1993. *Maya Cosmos: Three Thousand Years on the Shaman's Path*. New York: William Morrow.

Granovetter, M. 1979. "The Idea of 'Advancement' in Theories of Social Evolution and Development." *American Journal of Sociology* 85:489–515.

Hammer, M. 1983. "'Core' and 'Extended' Social Networks in Relation to Health and Illness." *Social Science Medicine* 17:405–11.

Hardin, G. 1968. "The Tragedy of the Commons." *Science* 162:1243–48.

———. 1991. "Paramount Positions in Ecological Economics." In *Ecological Economics*. R. Costanza, ed. New York: Columbia University Press.

López, A., S. Atran, J. Coley, D. Medin, and E. Smith. 1997. "The Tree of Life: Uni-

versals of Folkbiological Taxonomies and Inductions." *Cognitive Psychology* 32: 251–95.
Lundell, C. 1937. "The Vegetation of Petèn." Publication 219. Washington D.C.: Carnegie Institution of Washington.
Martin, F. 1992. *Common-pool Resources and Collective Action*. Bloomington: University of Indiana, Workshop in Political Theory.
Mitchell, J. 1969. *Social Networks in Urban Situations*. Manchester: Manchester University Press.
Nations, J. 1992. "Xateros, Chicleros, and Pimienteros: Harvesting Renewable Tropical Forest Resources in the Guatemalan Petén." In *Conservation of Neotropical Forests*. K. Redford and C. Padoch, eds. New York: Columbia University Press.
Nations, J., and R. Nigh. 1980. "Evolutionary Potential of Lacandon Maya Sustained-yield Tropical Forest Agriculture." *Journal of Anthropological Research* 36:1–30.
Ostrom, E. 1990. *Governing the Commons*. Cambridge: Cambridge University Press.
Rappaport, R. 1979. *Ecology, Meaning, and Religion*. Berkeley: North Atlantic Books.
Remmers, G., and E. Ucan Ek'. 1996. La roza-tumba-quema maya: Un sistema agroecológica tradicional frente el cambio tecnológico. *Etnoecológica* 3:97–109.
Rice, D., and S. Schwartz. 1992. Modern Agricultural Ecology in the Maya Lowlands. Paper presented at the 57th annual meeting of the Society for American Archaeology.
Romney, A. K., S. Weller, and W. Batchelder. 1986. "Culture as Consensus." *American Anthropologist* 88:313–38.
Schwartz, N. 1990. *Forest Society*. Philadelphia: University of Pennsylvania Press.
———. 1995. "An Anthropological View of the Forest Culture of Petén, Guatemala." Paper presented at the annual meeting of the American Association for the Advancement of Science, February 17, Atlanta, Ga.
Stewart, S. 1980. *Grammatica Kekchi*. Guatemala: Editorial Academica Centro Americana.
Wellman, B. 1979. "The Community Question: The Intimate Networks of East Yorkers." *American Journal of Sociology* 84:1201–31.
Wilson, R. 1995. *Maya Resurgence in Guatemala: Q'eqchi' Experiences*. Norman: University of Oklahoma Press.
Wisniewski, E., and D. Medin. 1994. "On the Interaction of Theory and Data in Concept Learning." *Cognitive Science* 18:221–81.
Wittgenstein, L. 1958. *Philosophical Investigations*. 2d ed. Oxford: Blackwell.

PART IV

Ethnoecology's Relevance

Local Knowledge in Global Context

12

Safeguarding Traditional Resource Rights of Indigenous Peoples

DARRELL A. POSEY

Safeguarding traditional knowledge and biogenetic resources has become a central issue in the expression of indigenous self-determination (IAITP 1992; Mead 1994; Posey 1994b; WCIP 1993). This concern has grown with increased awareness of the scale of past and present misappropriation of knowledge and resources by science, industry, and other commercial activities, and also because traditional resources are increasingly seen as the basis for greater political autonomy and economic self-sufficiency (Axt et al. 1993; Brush 1993; Gray 1991; Greaves 1994; Pinel and Evans 1994; Posey 1990). Intellectual property rights (IPR) have been proposed as a legal instrument under which indigenous peoples can seek protection for knowledge and resources. However, the inherent dangers lying within the IPR debate are well recognized by many indigenous peoples, who, along with many other researchers, think that IPR is not an appropriate mechanism for strengthening and empowering traditional and indigenous peoples (Bellagio Declaration 1993; Posey 1994a, RAFI 1993; Yamin and Posey 1993).

IPR developed as a Western concept to protect individual, technological, and industrial inventions. The term traditional resource rights (TRR) has emerged from the debate around IPR to reflect the necessity of reconceptualizing the limited and limiting concept of IPR (see Posey et al. 1996). This change reflects an attempt to build upon the concept of IPR protection and compensation, while recognizing that traditional resources—both tangible and intangible—are also covered under a significant number of other international agreements.

Biodiversity Prospecting and Economic Activities

To understand why the safeguarding of knowledge has recently become a major issue for indigenous peoples, consider the following points. First, the Earth Summit (United Nations Conference on Environment and Development), held in Rio de Janeiro in June 1992, was to a large extent about how biological diversity conservation could be economically exploited through biotechnological development, and it effectively highlighted the economic potential of traditional knowledge and resources. The Convention on Biological Diversity that emerged from the summit calls for the study, use, and application of "traditional knowledge, innovations, and practices." Its accompanying document, Agenda 21, actually outlines funding priorities to implement this process (Posey et al. 1995). As a result, considerable global funding will be directed toward the exploitation of indigenous knowledge and genetic resources.

Second, an increasingly large number of companies are "biodiversity prospecting"—that is, looking for biogenetic resources (plants, animals, bacteria, and so forth), including human genes, that can be used in the biotechnology industry (Joyce 1994; Pistorius and van Wijk 1993; Reid et al. 1993). Quinine and curare attest to the fact that this is not a recent phenomenon. Never before, however, have there been so many companies and collecting organizations interested in biogenetic resources that have been nurtured, protected, and even improved by indigenous peoples. The Guajajara people of Brazil use a plant called *Pilocarpus jaborandi* to treat glaucoma (Jacobs et al. 1990). Although Brazil earns twenty-five million dollars a year from exporting the plant, the Guajajara have suffered from debt peonage and slavery at the hands of agents of the company involved in the trade. Furthermore, *Pilocarpus* populations have nearly been wiped out by ravenous, unsustainable collecting practices.

Lastly, many indigenous communities need and are looking for economic alternatives. In the tropics, there are often few economic options other than timber extraction, mining, and ranching. Yet the tropical ecosystems are constantly touted as being one of the richest in biodiversity, with a huge potential for discoveries of new medicines, foods, dyes, fertilizers, essences, oils, and molecules of prime biotechnological use. In summary, the problem of knowledge and genetic resource exploitation now experienced by indigenous communities is only the start of a huge avalanche.

The Right to Say "No" and Categories of Protection

The first concern stated by indigenous peoples in every international forum is their right not to sell, commoditize, or have expropriated certain domains of knowledge and certain sacred places, plants, animals, and objects. Subsequent decisions to sell, commoditize, or privatize are only possible if this basic right can be exercised.

At least nine categories of traditional resources/indigenous intellectual property can be identified that a people or community may wish to protect (Posey 1994a):

1. Sacred property (images, sounds, knowledge, material culture, or anything that is deemed sacred and, thereby, not for commercialization);
2. Knowledge of current use, previous use, or potential use of plant and animal species, as well as soils and minerals, known to the cultural group;
3. Knowledge of preparation, processing, and storage of useful species;
4. Knowledge of formulations involving more than one ingredient;
5. Knowledge of individual species (planting methods, management practices, selection criteria, and so on);
6. Knowledge of ecosystem conservation (that protects the environment or may itself be of commercial value);
7. Genetic resources that originate (or originated) on indigenous lands and territories;
8. Cultural heritage (images, sounds, crafts, arts, performances); and
9. Classificatory systems of knowledge.

Quite clearly, knowledge is a common thread connecting all of these categories. Many indigenous groups have expressed their desire that all of these be protected as part of the larger need to protect land, territory, and resources and to implement self-determination (examples include the Matatuua Declaration, Kari-Oca Declaration, Declaration on Intellectual Property Rights and Indigenous Peoples). Control over cultural, scientific, and intellectual property is de facto self-determination, although only after rights to land and territory are secured by law and practice (that is, boundaries are recognized, protected, and guaranteed by law). But, as many indigenous peoples have discovered, even guaranteed demarcation of land and territory does not necessarily mean free access to the resources on that land or territory, nor even the right to exercise one's own culture or be compensated for the genetic resources that they have kept, conserved, managed, and molded for thousands of years.

The Search for an Alternative Framework

Significant efforts are being made to develop alternative legal frameworks to protect indigenous intellectual, cultural, and scientific property rights (Nijar 1994; Nijar and Ling 1993; UNDP 1994; TWN 1995).

A wide range of international agreements, declarations, and draft documents exist with relevance for building a newly designed system to protect traditional resource rights. These are labor law, human rights laws and agreements, economic and social agreements, intellectual property and plant variety protection, farmers' rights, environmental conventions and law, religious freedom acts, cultural property and cultural heritage, and customary law and traditional practice. Highlights from each of these areas are briefly described below.

Labor Law: IPR and the ILO

The International Labor Organization (ILO) was the first United Nations organization to deal with indigenous issues. In 1926 it established a Committee of Experts on Native Labor to develop international standards for the protection of native workers. In 1957 the ILO produced the Convention Concerning the Protection and Integration of Indigenous and Other Tribal and Semi-Tribal Populations in Independent Countries (Convention 107). Thirty years later this was rewritten as the Convention Concerning Indigenous Peoples in Independent Countries (Convention 169), with much of the original's "integrationist language" removed. The convention's key contribution is to guarantee indigenous peoples' rights to determine and control their own economic, social, and cultural development. The convention also recognizes the collective aspect of indigenous possessions, which is of obvious importance to IPR issues, since collectivity is fundamental to transmission, use, and protection of traditional knowledge. Until now, Convention 169 has not been sufficiently used with implementation of IPR in mind.

Human Rights and Intellectual Property

International human rights laws offer some mechanisms for cultural protection (Chapman 1994; Shelton 1995). The principal problem is that these are oriented toward nation-states and do not easily "provide a basis for claims against multinational companies or individuals who profit from traditional knowledge." The 1948 Universal Declaration of Human Rights and subsequent International Covenant on Economic, Social and Cultural Rights (1966) guarantee fundamental freedoms of personal integrity and action, political rights, social and economic rights, cultural rights, and equal protection under the law. Included here is the right of self-determination, including the right to dispose of natural wealth and resources. This also implies the right to protect and conserve resources, including intellectual property.

Significantly, they also protect the right to own collective property, as well as guaranteeing the right to just and favorable remuneration for work, which can be interpreted as work related to traditional knowledge. Finally, they provide for "recognition of interest in scientific production, including the right to the protection of the moral and material interests resulting from any scientific, literary or artistic production."

This language is echoed in the Draft Declaration on the Rights of Indigenous Peoples which states: "Indigenous peoples have the right to the protection and, where appropriate, the rehabilitation of the total environment and productive capacity of their lands and territories, and the right to adequate assistance including international cooperation to this end." It is clear that IPR should be seen as a basic human right worthy of incorporation in the campaigns of human rights organizations.

Economic and Social Agreements

In 1972, the United Nations Economic and Social Council formed a special human rights subcommission to study the problem of discrimination against indigenous peoples. After releasing a lengthy report that found inadequate protection of indigenous peoples' rights within existing international instruments, the subcommission released various resolutions recommending that the UN "provide explicitly for the role of indigenous peoples as resource users and managers, and for the protection of indigenous peoples' right to control their own traditional knowledge of ecosystems." They also requested that the secretary-general prepare a concise report on the extent to which existing international standards and mechanisms serve indigenous peoples in the protection of their intellectual property. The human rights commission has played an important role in pressuring other UN agencies to take action, through these calls for protection of, and protection for, indigenous peoples' IPR.

Folklore and Plant Variety Protection

The United Nations Educational, Scientific and Cultural Organization (UNESCO) should be a logical forum for IPR discussion, but, although UNESCO has heard "petitions" of complaints by native peoples related to the fields of education, science, culture, and information, indigenous questions remain marginal to its agenda.

The World Intellectual Property Organization (WIPO) in Geneva has 123 member states that have reached broad agreements on "industrial property" and "copyright." Within this framework, however, indigenous IPR, as collective property, would be considered folklore and not protectable. In 1984, however, UNESCO and WIPO, developed Model Provisions for National Laws on the Protection of Expressions of Folklore against Illicit Exploitation and Other Prejudicial Actions, which recognized individual and collective folklore traditions. Though never ratified, these provisions—backed up by criminal penalties—proposed protection of folklore, including material that has not been written down. The second important contribution was to provide for copyright protection of folkloric performances. Within WIPO's jurisdiction, the Union for the Protection of New Varieties of Plants provides protection to breeders of new plant varieties that are "clearly distinguishable," "sufficiently homogeneous," and "stable in essential characteristics."

The critical factor here is to link folklore and plant genetic resources with intellectual property. It is this complicated legal linkage that allows for expansion of the concept of IPR to include traditional knowledge not only about species use but also about species management. Thus, ecosystems that are molded or modified by human presence are a product of indigenous intellectual property as well, and, consequently, are products that are themselves protectable. Furthermore, "wild," "semi-domesticated" (or "semi-wild"), as well as domesticated plant and animal species are products of human activity and should also be protectable.

Farmers' Rights and the FAO

The UN Food and Agriculture Organization (FAO) has worked to find ways that will enable developing countries and "Third World farmers" to obtain a share in the huge global seed market. The questions of "farmers' rights" and "breeders' rights" have been extensively debated in this context (Crucible Group 1994; Fowler and Mooney 1990; GRAIN 1995a,b; Juma 1989; Kloppenburg 1988; Shiva 1994a,b). In 1987, FAO established a fund for plant genetic resources with the idea that commercial seed producers would voluntarily contribute according to the volume of their seed sales in order to finance projects for sustainable use of plant genetic resources in the Third World. Unfortunately, major seed producers such as the United States opposed mandatory contributions to the fund, which, in consequence, has been totally inadequate to accomplish its goal.

Environmental Law: Life after the Earth Summit

The Rio Declaration that emerged from the Earth Summit highlighted the central role of indigenous peoples in attaining sustainable development. The summit's legally binding Convention on Biological Diversity (CBD) does not explicitly recognize IPR for indigenous peoples, but its language can easily be interpreted to call for such protection. Following effective lobbying by indigenous organizations, signatories to the convention have pledged to respect, preserve, and maintain knowledge, innovations, and practices of indigenous and local communities embodying traditional lifestyles relevant for the conservation and sustainable use of biological diversity and to promote their wider application with the approval and involvement of the holders of such knowledge, innovations, and practices, as well as to encourage the equitable sharing of the benefits arising from the utilization of such knowledge, innovations, and practices. Agenda 21, which accompanies the Convention, specifically includes indigenous peoples and traditional knowledge in its "priorities for action" toward sustainable development.

Religious Freedom

In a seminar on IPR at the United Nations Human Rights Convention in Vienna in June 1993, Ray Apoaka of the North American Indian Congress suggested that IPR is essentially a question of religious freedom for indigenous peoples. "Much of what they want to commercialize is sacred to us. We see intellectual property as part of our culture—it cannot be separated into categories as [Western] lawyers would want." Pauline Tangipoa, a Maori leader, agrees: "Indigenous peoples do not limit their religions to buildings, but rather see the sacred in all life."

Cultural Property

In recent years, indigenous peoples have been increasingly successful in reclaiming the tangible aspects of their cultures, or "cultural property," from museums and institutions. This term has yet to be clearly defined, but it has come to refer to everything from objects of art to archaeological artifacts, traditional music and dance, and sacred sites. The concept of "cultural heritage" has appeared as a related "legal instrument" to link knowledge and information to the cultural artifact, and it has been used successfully as a legal tool in Australia.

Customary Law and Traditional Practice

During informal hearings for the 1992 Earth Summit in Rio de Janeiro, indigenous representatives pointed out several basic problems with the concepts of intellectual and cultural property:

1. The categories between cultural, intellectual, and physical property are not as distinct and mutually exclusive for indigenous peoples as in the Western legal system;
2. Knowledge is generally communally held, and, although some specialized knowledge may be held by certain ritual or society specialists (such as shamans), this does not give that specialist the right to privatize the communal heritage; and
3. Even if legal IPR regimes were to be implemented, most indigenous communities would not have the financial means to implement, enforce, or litigate them.

It was clear that under some circumstances commercialization of knowledge and plant genetic resources might be desirable, but the prime desire for indigenous peoples was an IPR regime that supports their right to say "No" to privatization and commercialization.

Indigenous delegates meeting in Rio de Janeiro produced the Kari-Oca Declaration and Indigenous Peoples' Earth Charter. In Clause 95 it states that "indigenous wisdom must be recognized and encouraged," but it warns in Clause 99 that "usurping of traditional medicines and knowledge from indigenous peoples should be considered a crime against peoples." Clause 102 of the Kari-Oca Declaration is explicit about indigenous peoples' concern regarding IPR issues: "As creators and carriers of civilizations that have given and continue to share knowledge, experience, and values with humanity, we require that our right to intellectual and cultural properties be guaranteed and that the mechanism for each implementation be in favor of our peoples and studied in depth and implemented. This respect must include the right over genetic resources, gene banks, biotechnology, and knowledge of biodiversity programs."

Since the Earth Summit, dozens of conferences, seminars, and workshops have been held by indigenous peoples to discuss the evolving IPR debate. During the 1993 UN Year for the World's Indigenous Peoples, intellectual and cultural property rights were on the agenda of nearly every major indigenous encounter. One of the areas of IPR research that is most lacking is that of non-Western IPR regimes. Up to now, the debate has centered around UN and Western concepts of intellectual and genetic property. But what about the property regimes of indigenous peoples themselves? A synthesis and analysis of non-Western systems would be very helpful in finding creative solutions to the problems of IPR protection. This research has yet to be carried out, and ethnobiologists are in a favorable position to contribute to this urgent and needed area.

Conclusion

It is fundamental that IPR/TRR should not be used to simply reduce traditional knowledge into Western legal and conceptual frameworks; indigenous legal systems and concepts of property rights should guide the debate. It will be indigenous and traditional peoples themselves who will define TRR in many different ways through practice and experimentation. The role of scientists, scholars, and lawyers should be to provide information and ideas.

The image of the "objective" scientist interested only in information for "purely scientific purposes" is, however, severely tarnished. Scientists and scientific institutions have become heavily involved—actively or passively—with the private sector. Plant, animal, and cultural material collected with public funds for scientific, nonprofit purposes are now open for commercial exploitation. Research, even in universities and museums, is increasingly funded by corporations, leaving questions of who controls the resulting data unanswered, even unasked. Scientific data banks have become the "mines" for "biodiversity prospecting." The publishing of information, traditionally the hallmark of academic success, has become a superhighway for transporting restricted (or even sacred) information into the unprotectable "public domain."

As a result, ethnoecologists are increasingly seen by indigenous, traditional, and local communities not as allies but as instruments of corporate interests. Scientists are not accustomed to playing the role of villain, and they find the lack of trust in their activities to be of profound puzzlement. They see themselves as seekers of the truth, and, if anything, victims of a system that exploits their efforts by low pay, reduced research funds, and precarious infrastructural support.

For ethnoecology to advance, it will have to find ways of developing equitable relationships with local communities in a way that benefits all parties.

TRR may very well provide the methodology and philosophical focus that can catalyze this process. Scientists, including most ethnobiologists and ethnoecologists, will have to undergo considerable reform to reach the point where genuine dialogue and negotiation can take place. This implies major changes in codes and standards of conduct developed by professional and scientific societies.[1] It also requires developing a cross-cultural scientific methodology that has never been perfected by any science, including anthropology.

Ethnoecology is in the enviable position of being interdisciplinary, with dedication to cross-cultural understanding. If the "new science" of dialogue cannot be developed out of ethnoecology, then where could it come from? Indeed, in view of increasing challenges to Western frameworks on intellectual property rights, ethnobiologists and ethnoecologists face the most daunting—and exciting—of challenges of any discipline looking toward the twenty-first century.

Note

1. The International Society for Ethnoecology has established a special committee to develop a Code of Ethics and Conduct based on a newly revised constitution. The drafts of these are being developed in conjunction with indigenous lawyers and will be debated in upcoming congresses in Nairobi (1996) and Aotearoa/New Zealand (1998). The Aotearoa Congress will be hosted and organized by the Maori Congress, and it is expected that the code and constitution will be adopted at that congress. See the *Bulletin of the Working Group on Traditional Resource Rights* (1996) 3:22–28.

References

Axt, J. R., M. L. Corn, M. Lee, and D. M. Ackerman. 1993. "Biotechnology, Indigenous Peoples, and Intellectual Property Rights." Washington D.C.: CRS Report for Congress, April 16.

Bellagio Declaration. 1993. "Statement of the Bellagio Conference: Cultural Agency/Cultural Authority: Politics and Poetics of Intellectual Property in the Post Colonial Era," Bellagio, Italy, March 11.

Brush, S. B. 1993. "Indigenous Knowledge of Biological Resources and Intellectual Property Rights: The Role of Anthropology." *American Anthropologist* 95(3):653–86.

Chapman, A. R. 1994. "Human Rights' Implications of Indigenous Peoples' Intellectual Property Rights." In *Intellectual Property Rights for Indigenous Peoples: A*

Sourcebook. T. Greaves, ed. Oklahoma City, Okla.: Society for Applied Anthropology.

Crucible Group. 1994. *People, Plants and Patents: The Impact of Intellectual Property on Biodiversity, Conservation, Trade, and Rural Society*. Ottawa: International Development Research Centre.

Cunningham, A. B. 1993. *Ethics, Ethnobiological Research, and Biodiversity*. Gland, Switzerland: WWF.

ECOSOC (UN Economic and Social Council, Commission on Human Rights, Sub-Commission on Prevention of Discrimination and Protection of Minorities. Working Group on Indigenous Populations). 1993a. "Discrimination against Indigenous Peoples: Study on the Protection of the Cultural and Intellectual Property of Indigenous Peoples."

———. 1993b. "The Mataatua Declaration on Cultural and Intellectual Property Rights of Indigenous Peoples," June.

FoE (Friends of the Earth). 1992. *The Rainforest Harvest: Sustainable Strategies for Saving the Tropical Forests: Proceedings of an International Conference Held at the Royal Geographical Society*. London: Friends of the Earth.

Fowler, C., and P. R. Mooney. 1990. *Shattering: Food, Politics and the Loss of Genetic Diversity*. Tucson: University of Arizona Press.

GRAIN. 1995a. "The International Technical Conference on Plant Genetic Resources: Opportunities for NGOS to Get Involved." *Biobriefing* No. 5.

———. 1995b. "The Green Revolution in the Red." *Seedling* 12(1):16–18.

Gray, A. 1991. *Between the Spice of Life and the Melting Pot: Biodiversity Conservation and Its Impact on Indigenous Peoples*. Document 70. Copenhagen: IWGIA.

Greaves, T., ed. 1994. *Intellectual Property Rights for Indigenous Peoples: A Sourcebook*. Oklahoma City, Okla.: Society for Applied Anthropology.

IAITP (International Alliance of the Indigenous-Tribal Peoples of the Tropical Forests). 1992. "Charter of the Indigenous-Tribal Peoples of the Tropical Forests." Penang, February 15.

Jacobs, J. W., C. Petroski, P. A. Friedman, and E. Simpson. 1990. "Characterization of the Anticoagulant Activities from a Brazilian Arrow Poison." *Thrombosis and Haemostasis* 63(1):31–35.

Johnson, M. 1992. *Lore: Capturing Traditional Environmental Knowledge*. Ottawa: Dene Cultural Institute/IDRC.

Joyce, C. 1994. *Earthly Goods: Medicine-Hunting in the Rainforest*. Boston: Little, Brown and Co.

Juma, C. 1989. *The Gene Hunters: Biotechnology and the Scramble for Seeds*. Princeton: Princeton University Press.

Kloppenburg, J. R. 1988. *First the Seed: The Political Economy of Plant Biotechnology 1492–2000*. Cambridge: Cambridge University Press.

Mead, A. T. P. 1994. "Misappropriation of Indigenous Knowledge: The Next Wave of Colonisation." *Otago Bioethics Report* 3(1):4–7.

Nijar, G. S. 1994. "Towards a Legal Framework for Protecting Biological Diversity and Community Intellectual Rights—A Third World Perspective." Third World Network discussion paper. Second session of the ICCBD, Nairobi, June 20–July 1.

Nijar, G. S., and C. Y. Ling. 1993. "Intellectual Property Rights: The Threat to Farmers and Biodiversity." *Third World Resurgence* 39:35–40.

Pinel, S. L., and M. J. Evans. 1994. "Tribal Sovereignty and the Control of Knowledge." Pp. 41–55 in *Intellectual Property Rights for Indigenous Peoples: A Sourcebook*. T. Greaves, ed. Oklahoma City, Okla.: Society for Applied Anthropology.

Pistorius, R., and van Wijk. 1993. "Commercializing Genetic Resources for Export." *Biotechnology and Development Monitor* No. 15:12–15.

Posey, D. A. 1990. "Intellectual Property Rights and Just Compensation for Indigenous Knowledge." *Anthropology Today* 6(4):13–16.

———. 1994a. "International Agreements and Intellectual Property Right Protection for Indigenous Peoples." In *Intellectual Property Rights for Indigenous Peoples: A Sourcebook*. T. Greaves, ed. Oklahoma City, Okla.: Society for Applied Anthropology.

———. 1994b. "Traditional Resource Rights (TRR): De Facto Self- determination for Indigenous Peoples." In *Voices of the Earth: Indigenous Peoples, New Partners and the Right to Self-determination in Practice*. L. van der Vlist, ed. Amsterdam: Netherlands Centre for Indigenous Peoples.

Posey, D. A., and G. Dutfield. 1996. *Beyond Intellectual Property Rights: Towards Traditional Resource Rights for Indigenous Peoples and Local Communities*. Ottawa: IDRC.

Posey, D. A., G. Dutfield, and K. Plenderleith. (1995). "Collaborative Research and Intellectual Property Rights." *Biodiversity and Conservation* 4(8):892–902.

Posey, D. A., assisted by G. Dutfield, K. Plenderleith, E. da Costa e Silva, and A. Argumedo. (1996). *Traditional Resource Rights: International Instruments for Protection and Compensation for Indigenous Peoples and Local Communities*. Gland, Switzerland: IUCN.

Principe, P. P., 1989. "The Economic Significance of Plants and Their Constituents as Drugs." In *Economic and Medicinal Plants Research*. Vol. 3 H. Wagner, H. Hikino, and N. R. Farnsworth, eds. London: Academic Press.

RAFI. 1993. "'Immortalizing' the (Good?) Samaritan: Patents, Indigenous Peoples and Human Genetic Diversity." *Rural Advancement Foundation International Communiqué*, April.

Reid, W. V., S. A. Laird, C. A. Meyer, R. Gamez, A. Sittenfeld, D. H. Janzen, M. A. Gollin, and C. Juma. 1993. *Biodiversity Prospecting: Using Genetic Resources for Sustainable Development*. Washington, D.C.: World Resources Institute.

Shelton, D. 1995. "Fair Play, Fair Pay: Strengthening Livelihood Systems through Compensation for Access to and Use of Traditional Knowledge and Biological Resources." Report prepared under the auspices of WWF-International, Gland, Switzerland.

Shiva, V. 1994a. "Freedom for Seed." *Resurgence* (March/April):36–39.

———. 1994b. "The Need for Sui Generis Rights." *Seedling* 12(1):11–15.

TWN (Third World Network). 1995. "Patents on Life, Intellectual Property and the Environment." A collection of TWN papers and articles. Briefings for the CSD Session no. 2. Penang: TWN.

UNDP (United Nations Development Programme). 1994. "Conserving Indigenous Knowledge: Integrating Two Systems of Innovation." A study by the Rural Advancement Foundation International, commissioned by the UNDP, New York.

WCIP (World Council of Indigenous Peoples). 1993. *Presumed Dead . . . But Still Useful as a Human By-product*. Ottawa: WCIP.

Yamin, F., and D. A. Posey. 1993. "Intellectual Property Rights, Indigenous Peoples and Biotechnology." *Review of European Community and International Environmental Law: Transfer of Biotechnology and Genetic Resources* 2(2):141–48.

13

A Practical Primer on Intellectual Property Rights in a Contemporary Ethnoecological Context

DAVID J. STEPHENSON, JR.

The knowledge of indigenous peoples about their local ecologies, the biological, botanical, geographical, geological, and hydrological attributes of those ecologies, and the "folk" expressions by indigenous peoples of their ways of life constitute a rich, rapidly disappearing bonanza of resources for Western scientific, technological, and business enterprises to exploit. The manner in which these resources are shared with the world will significantly shape the world's political economy for generations. Such indigenous knowledge and expressions—primarily because they are, by definition, "indigenous"—apparently defy classification under the traditional Western rubric of "intellectual property" and therefore are not readily protected under this Western legal convention (Brush 1993, 1994; Brush and Stabinsky 1996; Posey 1996; Posey and Dutfield 1996; Rubin and Fish 1994:45–48; Soleri et al. 1994:24–25). This has made such knowledge, resource identification, and expression ripe for exploitation by nonindigenous institutions without fair compensation to the indigenous peoples who have been the source for such knowledge and expression (Kloppenburg and Gonzales 1994:133; Posey 1996:1; Rubin and Fish 1994: 26–30; Stephenson 1996). Because of the rapidity with which this valuable, increasingly scarce supply of indigenous knowledge, resources, and expressions is being tapped, this apparent conceptual and legal incompatibility between Western intellectual property rights (IPR) and indigenous knowledge poses a critical, urgent dilemma: How can this exploitation be prevented if there are no legal means compatible with indigenous cultures available to stop such cultural exploitation?

In the face of this urgent dilemma, the concept of "intellectual property" persists and presents itself as so tantalizing and so comprehensive that its value as a pragmatic, organizing concept for policing the distribution of indigenous peoples' knowledge and expressions cannot be easily discarded (Axt 1993; Posey 1994:226; Rubin and Fish 1994:45). As Posey observes, "the deadly serious race . . . to conserve the cultural and biological diversity of the planet is on, . . . and the notion of intellectual property rights (IPR) seems to be one of the most interesting intellectual, legal, economic and political tools available to us at present" (Posey 1994:226). (Posey, however, recently has coauthored a volume titled *Beyond Intellectual Property* in which he suggests that the term "traditional resource rights" is more compatible than "IPR" for indigenous peoples and traditional societies [Posey and Dutfield 1996]).

Traditional Western legal tools used to protect intellectual property rights are themselves dramatically changing in order to adapt to modern technology and new geographic reaches (Benko 1987 and Weil and Snapper 1990, cited in Brush 1993:657; Dratler 1994:1–8; Rubin and Fish 1994). These adaptations may, in turn, create more open-ended and flexible IPR tools that are more suitable for indigenous peoples than those in place even as recently as twenty years ago. For example, legal protocols, such as computer software licensing agreements, developed in the last three decades to protect rapidly evolving, communally developed computer software may have utility for protecting traditional, evolving, communal indigenous property (Stephenson 1994; Vogel 1994:29).

Moreover, if indigenous peoples, or advocates for their rights, neglect to consider utilizing traditional intellectual property tools, such neglect may cause them to inadvertently overlook characteristics of indigenous knowledge, resources, and expressions that are intrinsically compatible with traditional IPR concepts. To cite one example, some researchers have argued that traditional nonindustrial farming crop selection is so fluid that the characteristic crops of such farmers cannot be defined in terms of their genetic composition, only in terms of management strategies (Brush et al. 1988). However, this attempt to abandon traditional categories that may lend themselves to traditional IPR plant variety protection conventions or even patents (Soleri et al. 1994:25) "neglects the effect of deliberate human selection, management strategies, and environmental factors on crop evolution, which is widely recognized by plant geneticists as the basis for FV's [folk crop varieties]" (Harlan

1992:127–28; Soleri and Cleveland 1993:209). Even if there are other impediments to invoking Western IPR protections offered by patents and plant breeders' rights, such as the difficulty of identifying discrete genotypes because of widespread traditional seed sharing, other IPR protections, such as protocols, contracts, trademarks, and certification marks may be amenable to regulating folk variety seeds (Rubin and Fish 1994; Soleri et al. 1994:25; see also Stephenson 1994:184–86).

One of the ironies of intellectual property in modern industrial societies is that it is often analogous to the words of the folk song: "You never know what you've got till it's gone." The legal definition of a form of intellectual property that is particularly pertinent to indigenous knowledge, trade secrets (see below), is simply any information that gives business a competitive edge. Similarly, other forms of intellectual property such as trade names and trademarks are valuable because of the intangible good will that is associated with them. Often businesses do not know what gives them a competitive edge or what the value of such intangible "good will" is until a competitor appropriates the information, name, or trademark (Dorr and Munch 1995:44). Even then, on average, after discovering losses caused by intellectual property piracy estimated in the hundreds of thousands of dollars per act of piracy, over 70 percent of the victim companies elect not to pursue legal action, and instead choose merely to absorb their losses (Wilson 1992:2).

Indigenous peoples are no different in these important respects from modern corporations. Indigenous peoples have knowledge and familiarity with local natural resources, and they develop creative expressions of their cultural experiences that can give them an important competitive commercial advantage in the world marketplace (Brush 1993:660). Often, like modern companies, they do not recognize how valuable such knowledge, resource familiarity, or creative expressions are until others use them for pecuniary gain (Brush 1993:660–61; Elisabetsky 1991:11–13). Like many modern business entities, indigenous peoples until recently have been reluctant to assert any legal rights that they might have for the taking of such knowledge (Posey 1996:1; Elisabetsky 1991:11–13). However, that does not imply that some form of legal action grounded in IPR is not possible for indigenous peoples, any more than should the failure of most modern businesses to pursue legal action for acts of intellectual piracy. In fact, to the contrary, indigenous peoples are beginning to assert their rights under the rubric of IPR (Posey 1996; Rubin and Fish 1994:46; Suagee 1994).

This paper provides an overview of some of the special kinds of legal questions that one might want to consider when evaluating the possibilities of invoking Western notions of IPR in contemporary indigenous contexts. This paper is not a survey of IPR tools in such contexts. Numerous thorough surveys have already been written (Brush 1993; Posey 1996; Posey and Dutfield 1996; Suagee 1994; see also Rubin and Fish 1994, which reviews IPR in these contexts but primarily engages in a legal analysis similar to the one in this paper). Rather, this paper engages in a skeleton legal analysis in order to provide guidelines for undertaking an IPR project in a contemporary indigenous setting. This paper does not suggest that traditional IPR tools must in every instance be invoked in such contexts. Nor should they necessarily be invoked in any instance. However, it is a premise of this paper, for the reasons stated in the foregoing discussion, that it is a mistake to abandon an IPR analysis in such contexts out of hand. Adaptations of Western IPR tools may be, and have been, effectively utilized by indigenous peoples under certain conditions (King 1994; Rubin and Fish 1994). Although not perfect, such tools may be more suitable than any other legal protection, and the rush to exploit indigenous peoples' resources will not always permit a wait for more perfect protection from unjust exploitation.

Such IPR tools, however, have evolved and will continue to evolve in response to the socioeconomic and political challenges that arise when indigenous peoples and Western scientists and entrepreneurs mix. Anthropologists can potentially have important roles to play in these contact situations so as to provide the ingredients for the shaping of appropriate IPR tools (Brush 1993: 667).

A Consideration of Appropriate IPR Tools

"Intellectual property" is usually defined as "intangible personal property in creations of the mind" (Dratler 1994:1–2). The most common legal embodiments of intellectual property are patents, copyrights, mask works (a relatively new form of protection in semiconductor chip design), trade secrets, trademarks, and legal actions to protect against unfair competition or to enforce similar proprietary rights (Dratler 1994:1–2—1–3). In 1970, a statutory form of plant variety protection was created to protect sexually reproduced plants that differs from plant patents, which protect asexually reproduced varieties of plants (Brush 1993:654; Dratler 1994:1–4n.7).

There are two general paradigms that all forms of intellectual property follow (Dratler 1994:1–8). The first is represented by copyrights, patents, mask works, and probably plant variety protection. This paradigm gives innovators strongly exclusive rights, for limited times, to exploit their innovations commercially, and thus its primary public purpose is to provide incentives for innovation (Dratler 1994:1–8). The second paradigm encompasses trademarks, trade secrets, and unfair competition.

The categories of intellectual property tools within the ambit of the second paradigm have other, more important, goals than merely the encouragement of innovation. These include protecting the public against confusion and deception from imitations, enhancing competition through comparison shopping, preserving investments in reputation and good will, and, generally, helping to avoid unfair and deceptive means of competition (Dratler 1994: 1–10). Trade secret law helps maintain standards of commercial ethics and promotes economic efficiency by minimizing the need for wasteful and inefficient practical measures to ensure secrecy (Dratler 1994:1–10). Trademark and trade secret protection are not nearly as absolute as copyright and patent protection; thus, they are much less likely to impair free competition in the marketplace (Dratler 1994:1–11).

Of these two paradigms, for reasons that are discussed in more detail below, the second is generally recognized as far more compatible with the traditional cultures of indigenous peoples. When suggestions are made concerning the potential utility of an adaptation of traditional IPR methods, the two tools most frequently cited as offering the best potential for the positive adaptation of IPR in indigenous contexts are variations of trade secret licenses and trademarks, including what have been referred to as "biodiversity prospecting" (or simply "bioprospecting") protocols or contracts (Posey 1996:32; Rubin and Fish 1994; Soleri et al. 1994:25; Stephenson 1994).

In an incisive, comprehensive consideration of conventional IPR protection, Brush discusses four major obstacles that programs to implement traditional Western types of intellectual property rights protections for indigenous knowledge will confront (1993:663). He labels these obstacles as (1) the general knowledge problem; (2) the group identity problem; (3) the legal status problem; and (4) the market problem (1993:663–67).

The general knowledge problem arises from the uneven distribution of specialized knowledge within a group (Brush 1993:663). However, this is certainly not a problem that is unique to indigenous societies. To the contrary, "folk" knowledge is probably more apt to be homogeneously distributed in an indigenous society than technical scientific knowledge subject to IPR protection is in most modern industrial societies.

The group identity problem "concerns the right of certain groups to claim exclusive control over knowledge and resources" (Brush 1993:663). The questions that arise are these: Which groups have legitimate claims to such control, and how are those group boundaries defined? This is truly a knotty, practical problem, as McGowan and Ukeinya poignantly document in their discussion of efforts to provide a means for fairly compensating local people in southeast Nigeria for the use of their traditional medicinal plants (1994:63). Despite these obstacles, however, they did develop a tentative compensation plan. They conclude that they could have benefited greatly by having had more time to consult with local residents in order to allow them to devise a compensation plan by engaging in a "collective search" (1994:66).

This kind of applied ethnographic endeavor is also proposed by Brush as a way to devise appropriate compensation methods (1993:667). Thus, this group identity problem is not irresolvable, but it does pose an urgent challenge to anthropologists, particularly those specializing in ethnoscience—for example, ethnoecology, ethnobiology, ethnobotany, and ethnomusicology—and ethnic group differentiation, more than to any other professionals, because such anthropologists are uniquely skilled in tracing divergent ethnically based epistemological domains and carving out group boundaries based on ethnic criteria (Barth 1969; Berlin 1992; Brush 1993:667; Williams 1989).

The legal status problem arises from the inferior political status of most indigenous peoples and minority ethnic groups within nation-states. Citing the Kurds of Iraq and the Mayan peasants of Guatemala as examples, Brush notes that groups "that control the greatest wealth of indigenous knowledge and biological resources may be the most marginalized by nation-states" (1993: 664). Even those groups that do enjoy a modicum of political status enjoy it only arbitrarily—for example, Native Americans in the United States and marginalized indigenous groups in China (Brush 1993:664). Indeed, even while striving to fashion a new regime for protecting their intellectual and cultural property rights through multilateral treaties, UN declarations, and other forms of positive and customary international human rights laws, indigenous peoples simultaneously recognize that the protection of their cultural and intellectual property is closely related to the realization of their most fundamental rights.

So, while it is true that inferior political status is a critical impediment to the protection of indigenous intellectual property rights, improving the political status of indigenous peoples is itself dependent on developing effective means for protecting cultural and intellectual property (Posey 1996:1–2; Suagee 1994:224). The two are mutually interdependent struggles. This is a variant of a theme introduced earlier in this paper: It would be a mistake to abandon IPR considerations out of hand for indigenous peoples merely because of their inferior political status. Rather, it is precisely because they often have an inferior political status that a search for effectively protecting their intellectual and cultural property is even more urgent.

The market problem that is discussed by Brush is a close cousin to the legal status problem (1993:665–67). He delineates how a Mexican state monopoly on the most economically important biological product in the early 1960s, *Dioscorea*, a source for steroids that was protected as a trade secret, collapsed when drug companies developed alternative sources between 1963 and 1976 (1993:665). As long as Mexico maintained its monopoly, indigenous peasants profited. But when the market collapsed, their income from this indigenous product virtually disappeared (1993:665).

Another aspect of the market problem is the disjunction between indigenous uses of resources and knowledge and the value of such resources and knowledge to nonindigenous peoples. For example, Brush notes that the rosy periwinkle of Madagascar is valued in the West for its antileukemia properties, but there is "no evidence that it is used in indigenous treatments of cancer" (Brush 1993:665). This disjunction theoretically makes it difficult to value the contribution of indigenous knowledge when the product is ultimately marketed (Brush 1993:665).

Again, however, while the market problem described by Brush poses a true obstacle, it is not insurmountable. Shaman Pharmaceuticals, and an independent monitoring organization, the Healing Forest Conservancy, are developing a method for compensating indigenous contributors in stages: short-term, medium-term, and long-term (King 1994; Moran 1994; Rubin and Fish 1994:29–30). The important implicit assumptions of these compensation formulae are that even Western drug companies themselves cannot accurately predict the marketability of various indigenous products; that the initial discovery of products nevertheless has a tangible, marketable value; that the value of the respective contributions of indigenous knowledge and Western science are interdependent; and that as Western pharmaceutical companies begin to develop tangible markets for indigenously inspired products, the indigenous societies can reap additional proportionate benefits. Shaman candidly admits that its compensation formulae are not perfect and need further refinement (King 1994:80), but the relative success of its pioneering venture offers encouraging tangible prospects for future endeavors by multinational companies that are philosophically committed to providing reciprocal benefits to indigenous peoples who provide vital intellectual and cultural sources for products that are successfully marketed throughout the world.

A Synthetic Model

A laudable, comprehensive, synthetic model for developing a collaborative, reciprocal relationship between a multinational company and indigenous peoples of the kind adumbrated by Shaman Pharmaceuticals and the Healing Forest Conservancy has been proposed by Posey (1994). This model, titled "A Covenant on Intellectual, Cultural, and Scientific Property: A Basic Code of Ethics and Conduct for Equitable Partnerships, and between Responsible Corporations, Scientists or Institutions and Indigenous Peoples," comprehensively defines the wide array of categories subject to protection, which include (but are not limited to) sacred property such as images, sounds, and material culture; knowledge of plants, animals, and minerals; conservation methods; crafts and performances; and classificatory systems of knowledge (pp. 246–47). The model lists basic principles to be observed by all partners to the enterprise, principles to be observed by the company, scientist, or institution, principles to be observed by the indigenous group, principles for independent monitors to follow, and the responsibilities of the partnership to the culture, community, society, environment, and region (pp. 247–51). This model simultaneously assumes that traditional IPR tools have some utility to such partnerships and also that they have to be uniquely adapted to the novel partnership situations created by the inevitably fundamentally different sociocultural orientations of the indigenous and nonindigenous partners (p. 227).

A Legal Analysis of Indigenous Property Based on an Adaptation of Two Traditional IPR Tools

In summary, although many commentators have been highly skeptical of efforts to apply traditional Western IPR constructs to indigenous peoples (Brush 1993; Greaves 1994; Posey and Dutfield 1996), their reservations are often guarded by suggestions that some adaptations of traditional IPR constructs may be fruitful (Posey 1994:237; Posey and Dutfield 1996; Soleri et al. 1994:25). When suggestions are made of the potential utility of an adaptation of traditional IPR methods, the two tools most frequently cited as offering the potential for positive adaptation are variations of trade secret licenses and trademarks (Posey 1996:32; Rubin and Fish 1994; Soleri et al. 1994:25; Stephenson 1994).

Copyrights and patents are almost universally considered to be too impractical or too one-sided (in favor of nonindigenous persons), or both, in indigenous contexts (Brush 1993:656; Seeger 1991; Suagee 1994:201; but see Rubin and Fish 1994:47–48 for a discussion of how to facilitate patent protection of indigenous knowledge via contractual stipulations).

The balance of this paper will discuss a methodology for evaluating the potential for successfully adapting the Western convention of licensing trade secrets and trademarks to the modern ethnoecological contexts created by collaborative efforts to fashion partnerships between representatives of indigenous peoples and Western scientists and entrepreneurs. In order to provide an ethnographic grounding for this methodology, it will be illustrated with reference to data collected by Soleri and Cleveland and others on the increasingly prized indigenous crop of Hopi blue corn (Soleri and Cleveland 1993; Whiting 1936, 1939).

Licensing Intellectual Property

"Simply defined, licensing is granting rights in property without transferring ownership of it" (Dratler 1994:1–2). A license is a form of contract, or agreement, regarding rights. There have historically been three broad types of licenses in intellectual property: (1) technology licenses, covering such things as patents, trade secrets, confidential information, and copyrights in technical material (for example, computer software); (2) publishing and entertainment licenses, covering creative properties such as books, plays, and music; and (3) trademark and merchandising licenses (Dratler 1994:1–5—1–6).

Merchandising licenses have become increasingly important as owners of well-known trademarks have begun to license their trademarks outside the commercial ventures that first brought them success, such as "Disney" tableware and "Coke" pants (Dratler 1994:1–6). Such merchandising licenses may have potential contemporary utility in protecting indigenous peoples from exploitation by preventing the use of native tribal names on merchandise that does not truly originate from such tribes or with their blessing (Stephenson 1994:186).

Licenses usually cover a "bundle" of combinations of complex intellectual property rights. Western lawyers cannot draft meaningful, enforceable licensing agreements between Western companies without "a keen sensitivity to the nature of the commercial, creative, or industrial property at issue and to customs and practices in the industry" (Dratler 1994:1–7). Therefore, it is equally imperative that lawyers drafting licensing agreements involving indigenous intellectual and cultural property rights have a keen sensitivity to the customs and practices of the indigenous peoples who are parties to the agreements. It would accordingly behoove them to employ the services of applied ethnographers or members of the indigenous societies involved to assist with the drafting process.

Trade Secrets and Their Potential Utility to Indigenous Peoples

Trade secrets are virtually ubiquitous in the contemporary world political economy (Dorr and Munch 1995:44). Examples of trade secrets include organic technology (including, for example, lipsticks, face creams, hair conditioners, soaps, scents, furniture polishes, recipes for foods and beverages, and pharmaceuticals); complex technology; business methods (including, for example, such varied methods as food manufacturing and preparation and methods of dance instruction); customer lists (including, for example, route information and the characteristics and breakdown of customer traits); business knowledge (including, for example, the names of key decision makers); and pending patent applications (Dorr and Munch 1995:46).

In the modern business world, "creativity in defining what is trade secret and, thus, what is protectable" is considered "smart business acumen" (Dorr and Munch 1995:46). Nevertheless, as stated above, it is common for persons possessing protectable trade secrets to fail to recognize that they do (Dorr and Munch 1995:46–47). Such ignorance can place businesses at a serious competitive disadvantage. Trade secrets are generally developed at substantial time and cost, and consequently the use of an existing trade secret by another generally eliminates the need for a large investment in time and materials. This savings is tangible and marketable and thus presents a potential means for assessing the value of indigenous knowledge and resources.

The potential utility to indigenous societies of utilizing a form of trade secret protection is not only that it may provide them with a means of obtaining compensation for the substantial savings in time and materials that Western multinational corporations derive from indigenous knowledge and resources but also that such resources are "perhaps the easiest and least expensive" form of intellectual property rights to protect (Dorr and Munch 1995:44).

Trade secret protection is of ancient origin, found to be of utility in simpler societies during simpler times, as reflected in roots that can be traced back at least as early as ancient Rome (Schiller 1930, cited in Dorr and Munch 1995:51).

Moreover, unlike other forms of intellectual property such as patents and copyrights, which fall under the umbrella of the "first paradigm" (see above), trade secret protection (such as trademark protection) may be of indefinite duration. For example, possibly the most famous trade secret in history, the ingredients of Coca-Cola, although now available in a widely circulated publication, have still not technically entered the public domain and therefore remain protectable as a trade secret (Dorr and Munch 1995:54; Pendergrast 1993).

The term "trade secret" is somewhat of a misnomer. As noted above, the original standard for identifying a trade secret was simply anything that gave one business a competitive advantage over another. Although the notion that any such nebulous thing must also be "secret" probably has always been implicit, it did not become explicitly part of American law until relatively recently. Even with the adoption of an explicit requirement that a trade secret be truly "secret," how secret it must be remains a question for the one who tries the facts. Moreover, many times "trade secrets" are composed of several elements that may each be common knowledge and in the public domain but which, when taken together, are not generally known and therefore qualify for protection. Indeed, even the most modern cases construing trade secrets recognize several characteristics of trade secrets that are potentially useful for indigenous peoples who have not had any historical reason to protect indigenous knowledge for commercial reasons: (1) the secret need not be actually used in a business; (2) the secret does not have to have a good track record—that is, it can be of economic "potential" only; and (3) the "secret" can be known by some outsiders so long as it is not "generally known" (Wilson 1992:4) or is information that can be pieced together only from a number of different sources (Marzouk and Parry 1994:33).

Many indigenous societies have long traditions of "secret" societies and knowledge that is confined to only a privileged few (Bok 1982:45–58; Soleri et al. 1994:27; Wolff 1950:345–48). Among the Southwest Pueblo Native Americans such as the Hopi, for example, one of the potential attractions to traditional intellectual property law is that it embodies a means for maintaining the "secret" attributes of traditional, often sacred, places, rituals, and knowledge (Soleri et al. 1994:27). Therefore, even though most indigenous peoples have not had any historical reasons to keep knowledge and resources secret for commercial reasons, the notion of secrecy is not foreign to them and, indeed, in many instances it is perfectly congruent with well-established traditions. Consequently, there may already be institutional mechanisms in place in many indigenous societies to engage in programmatic protection of the secrecy of indigenous knowledge and resources in a manner that is sufficient to satisfy the relatively rigorous, but not impractical, requirements imposed by Western trade secret law (Dorr and Munch 1995:44; Dratler 1994:1–10).

Trademarks and Their Potential Utility to Indigenous Societies

Trademark protection, like trade secret protection, falls traditionally within the second paradigm of intellectual property forms and thus shares many of the same attributes (Dratler 1994:1–10). The term "trademark" generally refers to "any word, name, symbol, or device or any combination thereof adopted and used . . . to identify and distinguish" goods or unique products from those sold by others, "and to indicate the source of goods, even if that source is unknown." Several ethnographers have recognized the potential utility, even the eventual necessity, of trademark protection for indigenous peoples, because the use of trademarks and variations of trademarks, such as certification marks (which certify region or other origin and which are referred to sometimes in international law as "appellations of origin") and collective marks (which can be used to certify membership in a cooperative, association, organization, or collective group) by indigenous peoples could potentially prevent, and may be necessary to prevent, nonindigenous persons from deceiving consumers regarding the true origins and sanctioning authorities of putatively indigenous products, services, performances, and creative works (Posey 1996:32; Posey and Dutfield 1996:84–87; Soleri et al. 1994:24; Stephenson 1994:186; see also Dorr and Munch 1995:103).

Assessing the Potential Commercial Value of Indigenous Knowledge and Resources for Licensing, Trademark, Certification Mark, or Collective Mark Purposes

The foregoing discussion has shown that any licensing agreement that concerns indigenous intellectual property must be grounded in a thorough understanding of the pertinent ethnoecological context. Similarly, ethnographic knowledge is vital to identifying group characteristics for the purpose of establishing collectively grounded, or ethnically based trademarks, certification marks, or collective marks. However, it is important to understand that the whole ethnoecological context includes not only identifying the indigenous knowledge, resources, or expressions that some identifiable group of indigenous people seeks to protect but also the potential or actual commercial value of such knowledge, resources, and expressions, so that a fair compensation formula can be integrated into any licensing agreements and so that the potential costs and benefits of establishing trademarks, and variations of trademarks, can be assessed. Therefore, the following kinds of investigations that are typically undertaken to assess the commercial value of intellectual property should be undertaken with modifications to fit the appropriate ethnoecological situation (Parr 1993:249–59):

The actual or projected profits of any nonindigenous entity known to be capitalizing or contemplating capitalizing on the subject indigenous knowledge, expressions, products, or any imitations thereof;
Actual or projected unit sales of any indigenous products, expressions, or imitations thereof;
A demographic breakdown of the potential market for any indigenous product, expression, or imitation thereof;
An evaluation of the selling price of any potential product, information, or expression;
The costs of manufacturing, selling, and delivering the products or transmitting the knowledge or expressions;
The business climate, pertinent business plans, and forecasts.
Feasibility studies concerning space, manpower, and technological know-how;
An assessment of the required return on investments to remain solvent;
Research and development budgets;
Identification of licensing agreements already in place, and potentially confusing trademarks;
Identification of competition;
A general review of the effects of the larger economic picture;
Investment reports regarding similar products in the industry;
The life cycle of the product, information, or expressions; and
A review of pertinent public information about the marketability of similar products, information, or expression.

An Illustration: The Potential for Developing Trademark and Trade Secret Licenses for Hopi Blue Corn

Until recently, within the United States, blue corn foods and seeds "were only known and available in a few areas of the Southwest" (Soleri and Cleveland 1993:222). Traditionally, blue corn has been very important as both a staple and ceremonial food crop for the Hopi (Soleri and Cleveland 1993:212, 222). Today, however, there is an exploding commercial market in blue corn products that has been accompanied by an increase in the international commercial availability of blue corn seeds (Soleri et al. 1994:23–24; Soleri and Cleveland 1993:222). Commercial, Western agribusiness has begun to penetrate and affect traditional Hopi communities (Pinel and Evans 1994:51; Soleri et al. 1994: 24; Soleri and Cleveland 1993:221). This penetration can have adverse effects, such as discouraging traditional food production, which, in turn, may destroy traditional Hopi culture and lead to general social disintegration (Soleri and Cleveland 1993:212), but it also offers the potential for new markets that will generate much-needed income and jobs for the Hopi (Pinel and Evans 1994: 51; Soleri et al. 1994:24; Soleri and Cleveland 1993:212).

Because the market in blue corn products has already been tapped, it is probably feasible to undertake a comprehensive market analysis of the kind suggested in the preceding section and place a tangible value on the profit potential a Hopi variety of blue corn might have in the international marketplace.

In addition, Hopi blue corn is grown from seed varieties and by farming technologies that may reasonably be considered indigenous to the Hopi, even though their earliest origins are probably in Mesoamerica, and even though the Hopi discovered mail order seed houses in Denver and Phoenix early in this century, because their crop varieties were "subsequently selected by the natural environment and people according to biophysical and sociocultural criteria" (Soleri and Cleveland 1993:211–12; Whiting 1936, 1939, cited in Soleri and Cleveland 1993:212). Thus, Hopi farmers theoretically may be able, under certain circumstances, to legitimately claim an ownership interest (in Western IPR terms) in such blue corn varieties and farming technologies for the purposes of licensing the seeds, the farming technology, or the products— "Hopi" blue corn meal—for a profit. They could theoretically seek trade secret, trademark, certification mark, and collective mark position.

It is also reasonable to anticipate, however, that their right to do so would be hotly contested by competitors who have previously penetrated the international marketplace with blue corn products, probably from other sources, and even allegedly obtained a federally registered trademark for such products (such as blue popcorn) as "Hopi Blue" without compensation to the Hopi, and continuously used it for more than five years—thus not only making it putatively "incontestable" under federal law but also posing a potential legal threat to any Hopi who begins to market products under the name Hopi (Dorr and Munch 1995:142; Soleri et al. 1994:24). This thus poses a classic example of an indigenous people who may not have realized what it had of value to the international market until it was gone—or taken.

Nevertheless, even if a competitor does indeed have an "incontestable" registered mark, the competitor's mark may still be legally attacked, and thereby protected for use by true Hopi, if it can be shown that the registration was obtained by fraud; the mark has been abandoned; the true source of the goods or services has been misrepresented; the mark is descriptive of the product itself, a name (such as the Hopi tribe) or the geographical origin of the product; a potential or alleged infringer (such as a true Hopi group) is actually a senior user of the mark; or the mark is being used to violate antitrust laws (Dorr and Munch 1995:142–43). The potential costs of undertaking a pertinent factual and legal investigation of these and related issues and of securing such trade secret, trademark, certification mark, or collective mark protection, possibly through very costly and time-consuming litigation, therefore would also have to be carefully weighed when deciding whether to pursue these potentially viable legal options.

Moreover, in order to effectively implement such licensing agreements, at least two of the problems posed by Brush—the general knowledge problem (that is, Should individuals or groups be entitled to grant such rights?) and the group identity problem (that is, If a group, which group?)—would have to be grappled with (Brush 1993:663). A thorough, action-oriented, ethnographic portrait should be obtained by any legal consultant in order to resolve these problems, and such a portrait may be difficult to obtain because of the Hopis' understandable reluctance to "spend much time talking to researchers" (Soleri and Cleveland 1993:207). Any such ethnographic portrait should probably also address the considerations raised by Posey in his model covenant (1994: 244–51), because the Hopi, like most indigenous societies, ultimately would probably enter only into licensing agreements that incorporate these considerations. In short, it would not be easy to implement trade secret and trademark protection via licensing agreements for Hopi blue corn products.

However, the obstacles to such implementation are primarily rooted in

Hopi ethnoecology, not in impediments inherent in these Western IPR tools. For example, some Western IPR tools, such as licensing agreements, certification marks, and collective marks, are flexible enough to protect the kind of knowledge that is usually regarded as typically indigenous—that is, knowledge that is communally owned or claimed and constantly evolving (Stephenson 1994). More specifically, Western trade secret, trademark, certification mark, and collective mark licensing tools are probably flexible enough to protect Hopi indigenous blue corn seed, technology, and products. The challenge to establishing IPR protection that remains is the ethnographic identification of persons or groups among the Hopi who have vested interests in such indigenous seed, technology, and products and who are interested in investing the time and resources that are necessary to legally protect these vested interests given the challenges to such protection that they are likely to face in the contemporary international political economy.

Should no such persons or groups emerge among the Hopi, however, it should not be inferred that the Hopi are somehow not ready for the twenty-first century. It would only mean that the Hopi are just like the more than 70 percent of the established U.S. companies who elect not to undertake legal action to protect their intellectual property after it is pirated (Soleri et al. 1994: 23; Wilson 1992:2).

Conclusion

The knowledge, resources, and creative expressions of indigenous peoples have made immeasurable contributions to Western science, technology, and art, and such indigenous contributions are both increasingly prized and increasingly scarce. Providing a fair measure of compensation to indigenous peoples for their knowledge, resources, and creative expressions may not only benefit indigenous peoples financially but also nurture their sociocultural and political integrity. Ironically, improving the survival chances of international cultural diversity also offers the best prospects for ensuring continual enrichment to Western science, technology, and art.

However, efforts to specifically develop enforceable legal mechanisms for providing compensation to indigenous peoples for the use of their knowledge, resources, and creative expressions are relatively recent and the subject of considerable experimentation, skepticism, and commentary (Greaves 1994). This paper has shown that traditional Western IPR tools should not be viewed as a forbidding forest that is inherently incompatible with these kinds of knowledge, resources, and creative expressions for which indigenous peoples seek compensation. Rather, they should be viewed as an array of different tools, with some, such as trade secret, trademark, certification mark, and collective mark licensing, being of potentially great utility for indigenous peoples, because such tools are flexible enough to be adapted to specific contemporary, ethnoecological contexts by a collaborative network of people who are sensitive to the unique sociocultural characteristics of each context. Such people might include applied ethnographers and ethnographically informed legal counsel, and representative cultural brokers from within the indigenous societies themselves.

References

Axt, J. R. 1993. *Biotechnology, Indigenous Peoples, and Intellectual Property Rights.* Congressional Research Service.

Barth, F. 1969. *Ethnic Groups and Boundaries: The Social Organization of Cultural Difference.* London: Allen & Unwin.

Benko, R. P. 1987. *Protecting Intellectual Property Rights: Issues and Controversies.* Washington, D.C.: American Enterprise Institute for Public Policy Research.

Berlin, B. 1992. *Ethnobiological Classification: Principles of Categorization.* Princeton: Princeton University Press.

Bok, S. 1982. *Secrets.* New York: Pantheon Books.

Brush, S. B. 1993. "Indigenous Knowledge of Biological Resources and Intellectual Property Right: The Role of Anthropology." *American Anthropologist* 95:653–86.

———. 1994. "A Non-Market Approach to Protecting Biological Resources." In *Intellectual Property Rights for Indigenous Peoples: A Sourcebook.* Tom Greaves, ed. Oklahoma City, Okla.: Society for Applied Anthropology.

Brush, S. B., et al. 1988. "Agricultural Development and Maize Diversity in Mexico." *Human Ecology* 16:307–28.

Brush, S. B., and D. Stabinsky, eds. 1996. *Valuing Local Knowledge: Indigenous People and Intellectual Property Rights.* New York: Island Press.

Dorr, R. C., and C. Munch. 1995. *Protecting Trade Secrets, Patents, Copyrights, and Trademarks.* New York: John Wiley & Sons.

Dratler, J. 1994. "Licensing of Intellectual Property." New York: Law Journal Seminars-Press.

Elisabetsky, R. E. 1991. *Intellectual Property Rights, R&D, Inventions, Technology Purchase, and Piracy in Economic Development: An International Comparative Study.* R. E. Evenson and G. Ranis, eds. Boulder: Westview Press.

Greaves, T. 1994. "IPR: A Current Survey." In *Intellectual Property Rights for Indigenous Peoples: A Sourcebook.* T. Greaves, ed. Oklahoma City, Okla.: Society for Applied Anthropology.

Harlan, Jack R. 1992. *Crops and Man*. 2d ed. Madison, Wisc.: American Society of Agronomy/Crop Science Society of America.
King, S. 1994. "Establishing Reciprocity: Biodiversity, Conservation and New Models for Cooperation between Forest-Dwelling Peoples and the Pharmaceutical Industry." In *Intellectual Property Rights for Indigenous Peoples: A Sourcebook*. T. Greaves, ed. Oklahoma City, Okla.: Society for Applied Anthropology.
Kloppenburg, J., and T. Gonzales. 1994. "Between State and Capital: NGOs as Allies of Indigenous Peoples." In *Intellectual Property Rights for Indigenous Peoples: A Sourcebook*. T. Greaves, ed. Oklahoma City, Okla.: Society for Applied Anthropology.
Marzouk, T. B., and T. M. Parry. 1994. *Counseling High Tech Clients*. Washington, D.C.: Marzouk and Parry.
McGowan, J., and I. Ukeinya. 1994. "Collecting Traditional Medicines in Nigeria: A Proposal for IPR Compensation." In *Intellectual Property Rights for Indigenous Peoples: A Sourcebook*. T. Greaves, ed. Oklahoma City, Okla.: Society for Applied Anthropology.
Moran, K. 1994. "Biocultural Diversity Conservation through the Healing Forest Conservancy." In *Intellectual Property Rights for Indigenous Peoples: A Sourcebook*. T. Greaves, ed. Oklahoma City, Okla.: Society for Applied Anthropology.
Parr, R. L. 1993. *Intellectual Property Infringement Damages: A Litigation Support Handbook*. New York: John Wiley & Sons.
Pendergrast, M. 1993. *For God, Country and Coca-Cola*. New York: Collier Books.
Pinel, S. L., and M. J. Evans. 1994. "Tribal Sovereignty and the Control of Knowledge." In *Intellectual Property Rights for Indigenous Peoples: A Sourcebook*. T. Greaves, ed. Oklahoma City, Okla.: Society for Applied Anthropology.
Posey, D. A. 1994. "International Agreements and Intellectual Property Right Protection for Indigenous Peoples." In *Intellectual Property Rights for Indigenous Peoples: A Sourcebook*. T. Greaves, ed. Oklahoma City, Okla.: Society for Applied Anthropology.
———. 1996. *Traditional Resource Rights: International Instruments for Protection and Compensation for Indigenous Peoples and Local Communities*. Cambridge: IUCN.
Posey, D. A., and G. Dutfield. 1996. *Beyond Intellectual Property*. Ottawa: International Development Research Centre.
Rubin, S. M., and S. C. Fish. 1994. "Biodiversity Prospecting: Using Innovative Contractual Provisions to Foster Ethnobotanical Knowledge, Technology, and Conservation." *Colorado Journal of International Environmental Law and Policy* 5:23–58.
Schiller, A. A. 1930. "Trade Secrets in the Roman Law: The *Actio Servi Corrupti*." *Columbia Law Review* 30:837–45.
Seeger, A. 1991. "Singing Other Peoples' Songs." *Cultural Survival Quarterly* 15:36–39.
Soleri, D., and D. Cleveland. 1993. "Hopi Crop Diversity and Change." *Journal of Ethnobiology* 13:203–31.
Soleri, D., et al. 1994. "Gifts from the Creator: Intellectual Property Rights and Folk

Crop Varieties." In *Intellectual Property Rights for Indigenous Peoples: A Sourcebook*. T. Greaves, ed. Oklahoma City, Okla.: Society for Applied Anthropology.

Stephenson, D. J., Jr. 1994. "A Legal Paradigm for Protecting Traditional Knowledge." In *Intellectual Property Rights for Indigenous Peoples: A Sourcebook*. T. Greaves, ed. Oklahoma City, Okla.: Society for Applied Anthropology.

———. 1996. "A Comment on Recent Developments in the Legal Protection of Traditional Resource Rights." *High Plains Applied Anthropologist* 16:114–21.

Suagee, D. 1994. "Human Rights and Cultural Heritage, Developments in the United Nations Working Group on Indigenous Populations." In *Intellectual Property Rights for Indigenous Peoples: A Sourcebook*. T. Greaves, ed. Oklahoma City, Okla.: Society for Applied Anthropology.

Vogel, J. H. 1994. *Genes for Sale: Privatization as a Conservation Policy*. New York: Oxford University Press.

Weil, V., and J. Snapper. 1990. *Owning Scientific and Technical Information*. New Brunswick, N.J.: Rutgers University Press.

Whiting, A. F. 1936. "Hopi Indian Agricultural: I. Background." *Museum of Northern Arizona Museum Notes* 8(10).

———. 1939. "Ethnobotany of the Hopi." Museum of Northern Arizona Bulletin no. 15. Northern Arizona Society of Science and Art, Flagstaff.

Williams, B. 1989. "A Class Act: Anthropology and the Race to Nation across Ethnic Terrain." In *Annual Review of Anthropology 1989*. B. J. Siegel, A. R. Beals, and S. A. Tyler, eds. Palo Alto: Annual Reviews.

Wilson, C. H. 1992. "High Technology Litigation: Piracy of Intellectual Property." Rocky Mountain Regional Conference on Computer Law. Denver: CLE International.

Wolff, K. H., ed. and trans. 1950. *The Sociology of Georg Simmel*. New York: Free Press.

14

Toward Compensation

Returning Benefits from Ethnobotanical Drug Discovery to Native Peoples

KATY MORAN

This paper will make three points about returning compensation for drug discovery to native communities:

1. Most of the world's remaining tropical biodiversity is in areas inhabited by indigenous, or native, cultures that are today the primary in situ caretakers of the planet's diversity;
2. Native cultures' knowledge of medicinal plant use, or "biocultural diversity," is valuable to drug discovery and requires compensation;
3. A method called "communal compensation" can be designed to return benefits from drug companies to native communities and pilot projects conducted to assess its viability. I will describe how this method was evolved in the context of the work done by the Healing Forest Conservancy.

Biocultural Diversity

Biological Diversity and Human Health

Biological diversity refers to the number and variety of the genes, species, populations, communities, and ecosystems that provide the basis for life on earth. Biodiversity fulfills aesthetic and spiritual needs for much of humanity and benefits human welfare directly. For example, tropical plant species provide the basis for today's medicines and hold the promise of supplying useful chemicals for tomorrow's therapeutics. Tropical species are valuable because, whereas in temperate climates winter kills many insect predators of plants,

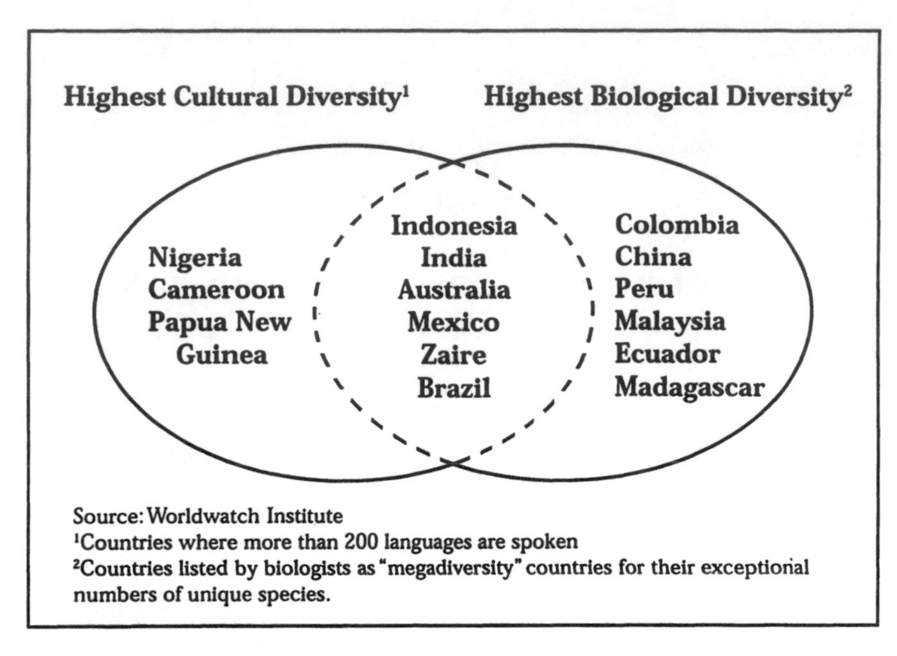

Figure 14.1. Cultural diversity and biological diversity circa 1990.

tropical species have no seasonal respite from insects. Consequently, they have evolved chemical protection from predators. The plant chemicals that have evolved to increase plant resistance against bacteria and other infectious organisms can also be therapeutically useful to humans.

However, if present trends in the loss of plant habitats in the tropics continue, it is predicted that as many as sixty thousand plants, nearly one in four of the planet's total, could be extinct by the middle of the next century. We remain unaware of promising cures we are losing, because, of the estimated 250,000 to 500,000 species of flowering plants alone, less than 1 percent have even been identified taxonomically. Fewer than half of those have been thoroughly investigated for their chemical composition and medicinal value (Farnsworth 1988).

Figure 14.1 shows where tropical biodiversity is located. Countries listed in the right circle and in the middle of the Venn diagram by Durning (1992) illustrate areas that biologists recognize as centers of biological diversity. These are called "megadiversity countries"—the twelve countries of the tropics with exceptional numbers of endemic or unique plant and animal species (that is, Colombia, China, Peru, Malaysia, Ecuador, Madagascar, Indonesia, India, Australia, Mexico, Zaire, and Brazil).

Cultural Diversity and Human Health

According to most anthropologists, the best single indicator of distinct cultures is spoken language, because it explicitly encapsulates the unique view of the universe held by the group that speaks the language. Of the six thousand languages spoken today, close to five thousand are languages of indigenous peoples. The Venn diagram shows countries, listed in the circle on the left and in the middle, that are the nine centers of cultural diversity. Countries are so defined by the number of languages spoken within each country (the nine are Papua New Guinea, Nigeria, Cameroon, Indonesia, India, Australia, Mexico, Zaire, and Brazil). As is evident, six of the twelve centers of biodiversity also rank among the top for cultural diversity (Indonesia, India, Australia, Mexico, Zaire, and Brazil). The relationship between biological and cultural diversity, or "biocultural diversity," is demonstrated by this diagram. Tropical ecosystems in which indigenous cultures live are where much of the biodiversity of our planet is also found.

"Indigenous peoples" is the term the United Nations uses, although in some countries "tribal" or "native" is the preferred term. According to the UN, there are 300 million indigenous peoples in about five thousand groups, spread around the world in more than seventy countries. Indigenous peoples are typically the descendants of original inhabitants of an area and are geographically isolated and distinct from dominant groups in language, culture, and religion. They describe themselves as custodians, not owners, of their land and resources, which are typically the base of their subsistence. Their social organization is tribal, referring to collective resource management and group decision making by consensus.

Indigenous peoples are not, as is sometimes romantically assumed, hands-off preservationists of nature. Humans constitute a species that uses nature for its own benefit, and indigenous cultures are no exception. But, typically, their local levels of biodiversity have not been lowered, as indigenous resource use does not deplete or concentrate pressure on a single biological resource as do resource-intensive systems such as monoculture. Rather, it distributes human impacts across the larger forest ecosystem. Within their habitats, indigenous peoples have laboratories of raw material from biotic resources to utilize for food, fiber, medicines, and other subsistence needs. Consequently, their traditional knowledge of the use of biological species represents libraries of information on the use of local biodiversity (Moran 1992).

This knowledge is embedded in indigenous peoples' cultural systems and intimately connected to their religious beliefs and cosmology (Reichel-Dolmatoff 1976). Plants are viewed as a gift of the gods who balance the universe, not commodities to be bought and sold by individuals in commercial markets. Accumulated over millennia and passed down generationally within communities, traditional knowledge of medicinal plants is as rich and diverse as are biotic resources, and as threatened. Since 1990, because of poverty, acculturation, outside encroachment, and loss of habitat, extinction has been the fate of an average of one indigenous culture each year in the Amazon region alone. But the loss of biodiversity and the traditional knowledge of plant use in the gene-rich developing world is a loss to all humanity, for future as well as present generations, particularly in regard to human health. For it is the knowledge of the local medicinal uses of tropical species from indigenous peoples that can guide the selection and screening of medicinal plants for use in drug discovery.

Use of Biocultural Diversity for Drug Discovery

Richard Schultes, the Harvard ethnobotanist widely regarded as the "father of ethnobotany," stated (1988):

> The accomplishments of aboriginal people in learning plant properties must be a result of a long and intimate association with, and utter dependence on, their ambient vegetation. This native knowledge warrants careful and critical attention on the part of modern scientific methods. If phytochemists must randomly investigate the constituents of biological effects of 80,000 species of Amazon plants, the task may never be finished. Concentrating first on those species that people have lived and experimented with for millennia offers a short-cut to the discovery of new medically or industrially useful compounds.

There are rich traditions in the use of medicinal plants on every continent. In the Americas, the use of herbals during the seventeenth and eighteenth centuries has been documented (Ford 1978), and the therapeutic use of plants in Africa is extensive today. In Asia, sophisticated systems involving plant-based medicines have been used for thousands of years, particularly in China and in the Ayurvedic system of India. According to the World Health Organization, 80 percent of the population of developing countries, about four billion people, depend on plant-based traditional medicines for their primary health care. Seventy-four percent of the 119 plant-based drugs used worldwide were discovered as a result of research to isolate their chemically active constituents. These include the antimalarial quinine found in *Cinchona*, the tranquilizer reserpine from the East Indian snakeroot, cardiac glycosides from digitalis, and the analgesics codeine and morphine, among many others (Farnsworth 1988).

Economic analysis by Artuso (1994) documents how indigenous knowledge adds value to plants screened for biological activity in the search for therapeutic chemicals. Rather than random screening, it is more efficient to use traditional knowledge of the medicinal use of plants as a lead for pinpointing promising plants. However, today there is no international legal instrument that specifically protects such intellectual property (Greaves 1994). Article 19 of the Draft Declaration of the UN Working Group on Indigenous Populations (1992) states that indigenous peoples have the right to protection, as intellectual property, for their traditional cultural manifestations, such as the medicinal knowledge of the useful properties of flora and fauna. Another pending document, the International Labor Organization's Convention 169, may also provide a more solid legal base for protection.

The Biodiversity Convention, which formalized the sovereignty of nations over their biodiversity, merely "encourages" equitable sharing of benefits arising from traditional knowledge, innovations, and practices. The convention does not, in its framework stage, establish mechanisms to operationalize and accomplish this equitability. It has not resolved, nor does it yet adequately acknowledge, the difference between ownership of biotic material by national governments and the ownership of the vital contribution of traditional medicinal use of the biotic material by indigenous peoples. Although equity is a critical issue in the spirit of the convention, during the 1994 meetings of the convention signatories in Nassau it was decided that issues concerning indigenous peoples' contribution to the drug discovery process would not be officially addressed until 1996. This delay occurred despite the numerous issues concerning indigenous peoples raised during various meetings.

To address these and other issues, during the last decade of the global resurgence of cultural identity, regional coalitions and federations of indigenous peoples worldwide have joined forces in a spirit of solidarity. The purpose of the federations is to strengthen the integrity of indigenous cultures. The coalitions are often not national because many indigenous groups are organized by ecosystems and dispersed across national borders. But they share a common ideology that is characterized by their relationship with the land. Their primary goal, as clearly stated in declarations that emerge from coalitions, is territorial rights to ancestral lands, including the right to limit access to indigenous territories. For example, in 1988 the Kuna Indians of Central America produced a manual for foreign scientists that describes Kuna terms for research undertaken in their territories. The manual describes the process for gaining permission to enter an area, establishes guidelines for research conduct, and encourages foreign scientists to transfer knowledge and technology to the Kuna during their research (Chapin 1991). Coalitions and federations comprise groups such as the Coordinating Body for the Indigenous Peoples' Organization of the Amazon Basin (COICA) in Amazonian countries, where more than two hundred tribes are located; the South and Mesoamerican Indian Information Center (SAIIC); and the World Rainforest Movement (Morris 1992).

Communal Compensation

In 1989, the Healing Forest Conservancy was founded as a nonprofit organization to work with federations to devise and implement a compensation strategy that promotes the conservation of tropical forests, particularly medicinal plants, and the conservation and welfare of tropical forest cultures, particularly their knowledge of medicinal plants (Moran 1994). Previously, no nongovernmental organization existed specifically to provide a formal and consistent process to compensate countries for the use of their biological diversity and communities for the use of their ethnobotanical leads to commercial development of biotic products.

The conservancy consults with indigenous federations, governments of tropical countries, professional associations of scientists, foundations, and other nonprofit organizations involved in the conservation of biocultural diversity. The focus of the conservancy is to develop communal compensation options in the form of programs with the following objectives:

1. To promote sustainable development by local harvesting of natural products in forests that might otherwise be cut for timber or cleared for farmland;
2. To generate local employment programs (where appropriate, focusing on women) by providing training in technical skills for species collection, identification, and inventory of local genetic resources by merging conventional and nonconventional scientific methods and processes;
3. To provide resources to survey, demarcate, and deed historic territories to indigenous groups;
4. To develop local markets for nontimber forest products such as medicinal plants;

5. To build and strengthen indigenous federations and institutions through education and communications between forest societies and the outside world; and
6. To link U.S. and international practitioners and policy makers in initiatives that foster the health and welfare of indigenous cultures and tropical forests.

The conservancy was founded through a donation from Shaman Pharmaceuticals, Inc., a natural products company in California focused on the discovery and development of novel pharmaceuticals from plants. As the name of the company implies, Shaman uses the science of ethnobotany, as well as isolation and natural products chemistry, medicine, and pharmacology, to create a more efficient drug discovery process. Although the young company has yet to commercialize a product, the use of ethnobotanic leads has brought potential products to clinical trials within a record time.

The traditional knowledge of plant resources for drug discovery is viewed by Shaman as the primary source for leads in drug discovery, and agreements with communities and countries that the company works with secure compensation for traditional knowledge and for medicinal plants when products are commercialized (Conte 1994). Converting the Biodiversity Convention into praxis, Shaman pioneered a novel concept for compensation (Artuso 1994). The company will return a percentage of product profits on a commercialized product to all of the indigenous communities and countries with which it has worked, regardless of where the actual plant sample or traditional knowledge originated (King 1994). Through this process, the risk of not receiving compensation for a commercialized product for individual countries or communities that contributed to product discovery is lessened. In a financially unpredictable industry, spreading the benefits and risks among all Shaman collaborators increases opportunities for compensation and hastens compensation returns. All Shaman collaborators benefit equally from risk-sharing provisions, acknowledging the spirit of the Biodiversity Convention and the need to separate the ownership of biotic species from the ownership of traditional knowledge.

After a product is commercialized, Shaman will channel a percentage of product profits for compensation through the conservancy. The conservancy will deliver "communal compensation" programs to Shaman collaborators that benefit the community, rather than cash payments to an individual, following indigenous systems of resource use.

Pilot Projects

Meanwhile, the conservancy has successfully sought support for compensation pilot projects from foundations, nonprofit organizations, and environmental organizations. Funders feel compensation programs have merit as stand-alone conservation programs also. Pilot programs were developed through input from three sources:

1. Indigenous federations and their declarations;
2. International conventions signed by governments of species-rich countries; and
3. Professional associations, such as the International Society of Ethnobiology, whose members collaborate with governments and federations.

Pilot projects are conducted to test the viability of compensation processes and to gain knowledge from practical experience learned during projects. Results are shared with communities and countries involved in the sustainable development of their biodiversity. They can then relate experiences from pilot projects to their specific circumstances and make an informed choice among compensation options.

Terra Nova

The government of Belize has designated a three-thousand-acre tract of land, named Terra Nova, to establish what many believe to be the world's first ethnobiomedical plant reserve. The land, located in the Cayo District of Belize, has terrain ideal for many types of Maya medicinal plants. It is also a nursery for "orphaned" species—medicinal plants integral to the survival of traditional healing in Central America that have been rescued from development project bulldozers and transplanted to a safe location at Terra Nova. To halt current poaching and logging near Terra Nova, the conservancy and the Rex Foundation have supported the surveying, demarcating, and deeding of the land, following a request to do so from the Belize Association of Traditional Healers (BATH). Title for management of Terra Nova has been deeded to BATH following an indigenous pattern of communally owned and managed resources. The association holds monthly public education meetings for local villagers at which traditional healers discuss medicinal plants and traditional remedies for a variety of common ailments. Future plans, supported by the conservancy and the New York Botanical Gardens' Institute of Economic Botany, include making Terra Nova a self-supporting extractive reserve through a factory of herbal remedies using Maya medicine. Included in the plan is a clinic with a sliding fee scale to help meet the health needs of nearby villagers whose primary health care is supplied by traditional Maya medicinals. Today, Terra Nova offers a way for traditional healers to train and pass on the healing forest legacy to future generations of practitioners in a reserve that unites traditional peoples' use of plants with sustainable extraction.

Future pilot projects include testing how indigenous groups can use low-cost, satellite-generated, global positioning systems (GPS) for the demarcation of their lands. Now available to remote communities, such technologies enable local groups to reference land boundary data and locate boundaries to an accuracy of between a few centimeters and twenty-five meters. Once referenced this way, data can be entered into a global environmental database such as the Global Environmental Monitoring System (GEMS), the Global Resource Information Database (GRID), or the Earth Observing System (EOS). Geographic Information Systems computer software systems can also be used by indigenous groups for natural resource assessments, conservation analysis, and land-use planning.

Medicine Woman

With support in the form of grants, the Healing Forest Conservancy has established an education and training program for women, particularly indigenous women, in tropical countries. Training programs are held to increase the informational value to biotic resources locally. The program teaches technical skills such as biological inventorying, chemical screening or bioassays, and ethnobotany to women from societies where the program suits local needs.

In 1994, the Asia Foundation sponsored a grant to the Healing Forest Conservancy to participate in the Environmental Fellowship Program, a component of the U.S.-Asia Environmental Partnership of USAID. As part of the fellowship, the conservancy funded a Medicine Woman pilot project in Lucknow, Uttar Pradesh, India, for students from various regions of the country. The pilot project was conducted in India because it was the site of the fourth International Congress of Ethnobiology, which the students also attended. India is rich in biocultural diversity, being home to over four hundred unique ethnic groups, 75 percent of which are tribal, and forty-five thousand species of plants, 30 percent of which are higher plants. The purpose of the course was to increase and diffuse the knowledge of ethnobiology. Areas such as basic technical skills in ethnobiology and informed participation in applied ethnobiology were explored through discussions on ethical issues and networking for students. A professional dialogue with other tribals and those associated with ethnobiology worldwide was also pursued.

The conservancy, in cooperation with the Asia Foundation, the Rex Foundation, and the National Botanical Research Institute of India, supported participation by twenty-eight trainees, half of whom were tribal women. Mornings were spent in classes discussing methods in ethnobiology with varied local as well as international specialists in the natural and social sciences. Discussions on topics such as medical anthropology, linguistics, and ethnomedicine were presented. Afternoons were spent in the field for technical training in collection methods such as how to make and use a plant press and methods of plant drying. Because collecting is labor intensive, training enhances employment opportunities with national governments that are responsible for completing country inventories. While training can generate employment by outsiders interested in developing bioresources commercially, participation in the collection process also supplies communities with greater control over the use of their resources. Moreover, regional medical needs can be addressed by ensuring the status and continuity of plant resources that provide health care through traditional medicines and viable plant reserves where future healers can be trained.

During a visit to the national herbarium, procedures such as the care and storage of dried plants and simple herbarium laboratory techniques were demonstrated. Discussions focused on ethical issues in applied ethnobiology, such as intellectual property rights. Students petitioned for more opportunities to add value to biotic resources in their own countries and communities (Akerele et al. 1991; Soejarto and Gyllenhaal 1992). They now recognize that information on plant and animal species, in itself, is valuable, and want skills to add informational value locally. For example, with training, field methods for performing simple tests to detect the biological and pharmacological activity of plants can be implemented locally. Currently, little of the chemical analysis of medicinal plants used by indigenous peoples is completed in situ, despite the advantages in analyzing fresh plant material. When plant material is air dried so that it can be analyzed later in a laboratory its chemistry changes, and some volatile compounds can be lost. Likewise, the separate analysis of individual plants normally taken together in a concoction ignores the chemical reactions that occur when chemicals from different plants are mixed (Prance 1991).

Training can build local capacity and supply income-producing opportunities to the communities that are most strategically located to collect, identify, analyze, and protect species. Inventories that describe local use can be undertaken in the local language to protect the information from outsiders and to leave a documentation of cultural knowledge within the community. More important, if training, education, and employment opportunities are available to women, ecological impacts are actually doubled, because women who have the option of choosing between having a job or having another child typically choose the job, effectively lessening population pressure on natural resources. Also, women have specialized, gender-related knowledge about plant use for contraception, abortifacients, nutrition, and infant and child care. Yet few programs offer this opportunity specifically to women.

National governments benefit from training programs by gaining a technological infrastructure for science and commerce that yields jobs and taxes. Fees can be charged to outsiders with commercial or research interests in the sustainable development of biotic resources, allowing debt-ridden nations to forgo short-term profits from logging, cattle grazing, and monoculture development projects that destroy forests. For example, new technologies to screen plants supply information that can lead to the design of new drugs. These technologies enhance the capacity of countries and communities to increase the value of their biotic resources. The pharmaceutical industry in biodiversity-poor countries currently prices plant collecting at $50 to $100 per raw plant sample, but that price doubles for a chemical extract. Countries can choose whether to supply natural products in the form of extracts, rather than raw, unprocessed material, to foreign investors, or to establish their own pharmaceutical industry. Adding value locally lowers the total cost of drug discovery from natural products in a high-risk, high-gain industry valued at $100 billion annually.

On the average, some ten thousand samples must be screened to yield one marketable drug. The whole process typically takes from twelve to fifteen years, at a cost of close to $300 million; however, there are several levels of entry for local communities into the sustainable development of biodiversity. These include plant collecting, taking inventories, conducting bioassays, recollection, harvesting, and herbarium storage and documentation. Training increases the capacity of biodiversity-rich countries to perform these services and eases their entry into the field of natural products, should they choose to do so. Proprietary arrangements can ensure that compensation from the commercialization of natural compounds or synthesized chemicals reverts equitably to the peoples and nations of original production.

Evaluations by students at the end of the Medicine Woman course in India were positive; they suggested future training courses of greater length to broaden their knowledge and experience. In 1995 the conservancy helped support another training course at the Limbe Botanic Garden in Cameroon organized by the Biodiversity Development and Conservation Programme, an African NGO (Moran 1997). In addition, an organization of Maya women of highland Chiapas has requested training in conducting an inventory of plant species in their language. Among other things, they will use the information to assess the nutritional status of the Maya diet.

Conclusions

Today, the sustainable development of biotic resources in the tropics is a work in progress, euphemistically called "an emerging science." Toward this end, conservancy pilot projects have generated important questions:

1. Can natural products be harvested over time while maintaining the integrity of the forest ecosystem and forest cultures?
2. Can local capacity be well enough organized to monitor and manage the process?
3. Do options such as land demarcation and training opportunities offer adequate and secure compensation to communities and countries and benefit future generations?

It is useful to pose economic and environmental questions on how to decrease the degradation of tropical forests and tropical forest peoples and the resultant loss of biocultural diversity. What is certain, however, is that equity stands out as a critical component of the discussion. For equity not only means compensation but, more important, it also means equal standing among participants in making decisions about what form compensation should take.

There is much faith today in the ability of market forces to save the rain forest. But market forces can be a double-edged sword when Western economic concepts are introduced to groups of forest peoples. However, this should never be an excuse for excluding indigenous groups from the process, for this is their, and only their, decision to make. How can we expect support for biocultural diversity conservation if it is perceived as "us" trying to prevent "them" from achieving what we already have? We should support indigenous efforts for development in a manner that unifies and strengthens communities through the use of traditional or compatible institutions and federations. If we are serious about the sustainable development of biodiversity for the discovery of useful therapeutics, participation and equal standing among all concerned must be the principle that guides the effort.

References

Akerele, O., V. Heywood, and H. Synge, eds. 1991. *The Conservation of Medicinal Plants*. Cambridge: Cambridge University Press.

Artuso, A. 1994. "Economic Analysis of Biodiversity as a Source of Pharmaceuticals." PAHO/IICA Conference on Biodiversity, Biotechnology and Sustainable Development. San Jose, Costa Rica, April 12–14.

Chapin, M. 1991. "How the Kuna Keep Scientists in Line." *Cultural Survival Quarterly* (Summer): 15–17.

Conte, L. 1994. "Testimony for Hearing on U.S. Ratification of the Convention on Biological Diversity." U.S. Congress, Senate Foreign Relations Committee. Washington, D.C.

Durning, A. 1992. *Guardians of the Forest*. Washington, D.C.: Worldwatch.

Farnsworth, N. R. 1988. "Screening Plants for New Medicines." In *Biodiversity*. E. O. Wilson, ed. Washington, D.C.: National Academy Press.

Ford, R. I. 1978. "The Nature and Status of Ethnobotany." Anthropological papers, no. 67. Ann Arbor: Museum of Anthropology, University of Michigan.

Greaves, T. 1994. *Intellectual Property Rights for Indigenous Peoples*. Oklahoma City, Okla.: Society for Applied Anthropology.

King, S. R. 1994. "Establishing Reciprocity: Biodiversity, Conservation and New Models for Cooperation between Forest-dwelling Peoples and the Pharmaceutical Industry." In *Intellectual Property Rights for Indigenous Peoples: A Sourcebook*. T. Greaves, ed. Oklahoma City, Okla.: Society for Applied Anthropology.

Moran, K. 1992. "Ethnobiology and U.S. Policy." In *Sustainable Harvest and Marketing of Non-timber Forest Products*. M. Plotkin and L. Famolare, eds. Washington, D.C.: Island Press.

———. 1994. "Biocultural Diversity Conservation through the Healing Forest Conservancy." In *Intellectual Property Rights for Indigenous Peoples: A Sourcebook*. T. Greaves, ed. Oklahoma City, Okla.: Society for Applied Anthropology.

———. 1997. "Returning Benefits from Ethnobotanical Drug Discovery to Native Communities." In *Biodiversity and Human Health*. F. Grifo and J. Rosenthal, eds. Washington, D.C.: Island Press.

Morris, K. 1992. *International Directory and Resource Guide*. Oakland, Calif.: South and Meso American Indian Information Center.

Prance, G. T. 1991. "What Is Ethnobotany Today?" *Journal of Ethnopharmacology* 32: 209–16.

Reichel-Dolmatoff, G. 1976. "Cosmology as Ecological Analysis: A View from the Rainforest." *Man: A Journal of the Royal Anthropological Institute* (11)3:307–18.

Schultes, R. E. 1988. "Primitive Plant Lore and Modern Conservation." *Orion Nature Quarterly* (7)3:8–15.
Soejarto, D., and C. Gyllenhaal, eds. 1992. "The Declaration of Belem and the Kunming Action Plan." *International Traditional Medicine Newsletter* 4:1.
United Nations Working Group on Indigenous Peoples. 1992. "Report on the Intellectual Property Rights of Indigenous Peoples," July 6.

15

Am I My Brother's Keeper?

CHRISTINE S. KABUYE

Ethnoecology's importance in environmental management and resource conservation is gaining recognition. At a time when natural resources are "disappearing" under the various processes of development and failing management practices, it has been found advantageous to turn to indigenous knowledge for possible solutions. Some concerns are focused on medicinal plants and some on food plants, while others are on promising natural products. It has been possible to document the rich heritage of ethnoecological knowledge only because some indigenous people or local communities, despite discouragement and sometimes ridicule, persisted against the odds and faithfully practiced and transmitted what they believed to be right. In other cultures, though, it has not been that simple, as the people themselves are threatened, along with their knowledge and resources.

The title for this paper is taken from the biblical Cain who, after killing his brother, Abel, was asked by God where his brother was. He retorted, "Am I my brother's keeper?" One wonders how many people would say the same thing concerning our responsibility to each other at the local and international level. If the earth's resources are to continue being available for the benefit of all, it is more than essential for each one of us to marshal our efforts in the same direction, each human being responsible for the other.

People's Relationship with Their Environment and Resources

People have depended on their environment and resources for survival, and their activities have affected habitats in various ways over time, whether on a small scale as hunter-gatherers or on a large scale as extensive and intensive agriculturalists and industrialists. As hunter-gatherers, people roamed the forest and woodland, leaving tracks and markers to enable them to return to the same trees, shrubs, and herbs to gather fruits and medicines, tending plants in the wild, setting traps, and collecting honey from bee hives or even gathering insects for food. Later, the domestication of some plants and animals led to mixed subsistence farming while hunting and gathering were still pursued. The management of resources was monitored and controlled in various ways. For instance, the herbalist would regulate the harvest of bark or root from one side of a tree at a time so as not to kill it, thereby ensuring a continued supply. The pastoralist managed grazing land by seasonal movement of livestock to dry-season grazing sites and the seasonal burning of grass to stimulate new growth. Similarly, the shifting agriculture of the peasant farmer allowed enough fallow period for recovery of soil fertility.

Allowing for periods of experimentation and innovation, these processes evolved over a long period of time, with the resultant relationships becoming deeply rooted. With such strong attachment to resources, the practices and norms prescribed and followed for their maintenance are embodied in people's cultures. Societies over time developed systems to ensure continued availability of, and access to, resources. One such system is the totemism connected with clans. This belief system is very old, at least in many African societies. In this system, members of a clan will not hunt or harvest their totem animal or plant. This provides a certain degree of protection to specific animals and plants. In addition, some trees, patches, or whole forests are deemed sacred. Sacred sites serve as places of worship, burial grounds, or places for communication with ancestors or for performing particular rituals. The resources here are protected by elders and may not be hunted, cut, or damaged in any way except under special circumstances. The sites may contain some medicinal plants, so in effect access to these valuable resources is regulated. People's daily lives are thus closely interwoven with nature, and the prevailing ethic, since nature provides for their livelihood, is that they have collective responsibility to nature for their own survival. Unfortunately, this relationship has tended to be somewhat threatened by changes in lifestyles and population pressure, which have led to the overexploitation of resources. As a consequence, unsustainable levels of resource extraction plague certain areas.

Indigenous Knowledge

Through their relationship with the environment and resources, communities acquire a body of knowledge often referred to as indigenous, local, or traditional. This is either collective and general or highly specialized and vested in particular members of society. Examples of the latter include the traditional healer with specialized knowledge of medicinals and the peasant farmer with special capability in selecting planting material or matching crops to soils. Because of increasing recognition of this fact, indigenous knowledge (IK) has recently become very important at global forums as a key strategy for sustainable development. IK includes cultural perceptions of nature, the environment, and natural processes. These perceptions guide decision making on practices for the management, maintenance, and utilization of the environment and natural resources. Most societies share their knowledge either through individual contact or during ceremonies. Unfortunately, because of the influence of the colonial era, IK in some areas was relegated to the backseat and dismissed as sheer superstition. The young, especially in urban areas, do not know or care much about it. Even in the villages, where the elderly used to pass on their knowledge to the young, intergenerational transfer has not been possible because the young are not particularly interested or available. Moreover, most IK is not recorded but preserved only in oral traditions.

There is thus a likelihood of losing the knowledge altogether unless remedial steps are taken to revive it. This creates a need to institute special research and training programs before it is too late. There is, for instance, an initiative in a small Maasai community in Kenya to have elders talk to children in schools about how they used to protect the environment. This is a potentially effective way of trying to pass on the information to the next generation. A system made up of programs like this would go a long way toward understanding, preserving, and using IK in development.

Challenges Created by Development

People's dependency on and harmony with their environment and resources have faced a number of challenges in the last century, a few of which are mentioned here.

Protected Areas

The establishment of national parks in many African countries and elsewhere has displaced people from the land and denied them access to resources they had considered their own. Although it was done for conservation's sake, the process has been extremely disruptive. The denial of access to resources has made people develop an intense dislike for national parks. Recently, park management instituted a benefit-sharing system in Kenya whereby people surrounding the parks receive part of the park entrance fees, but that is not the same as being an integral part of the reserve. Indigenous people have lost access to resources such as medicinal plants and, along with it, their spiritual attachment to the area. As protected areas go, there is a lot to recommend the Biosphere Reserve System, which allows local people access to resources.

Consumer Economies

Modernization and consumerism have continued to drain resources to unsustainable levels. Rapid industrial development has led many Western countries to lose their wild resources, and this could easily happen to developing countries unless such development is held in check. While developed countries are banking on the developing countries to conserve their biodiversity, which would also serve as a carbon sink to reduce global warming, certain consumptive models and requirements in the West are likely to have the opposite effect. The example of elegant courtrooms being built in the United States with planks of mahogany from Africa is a case in point. One wonders why it must be mahogany, and how we can talk of carbon sinks if mahogany, which is a major component of tropical forests, is extracted for such trivial purposes.

Access to Genetic Resources

For some time, there has been a bone of contention over folk crop varieties, or landraces, obtained from local communities and peasant farmers. Examples are local crop varieties of grains and tubers developed and maintained through many generations by peasant farmers. These were freely accessed for conservation, research, and improvement purposes from the local farmers, but with slight strain development by plant breeders they became patentable plant varieties with intellectual property rights attached. This was done without recognition of the part played by the peasant farmers in nurturing and breeding the folk varieties. The long process of domestication, selection, and maintenance—the contribution of farmers—cannot be ignored, as without that process the breeders would have nothing—no raw materials—with which to work. There has to be a mechanism by which the farmers' innovation and experimentation can be recognized as part of the crop development process. In any case there should be due recognition and compensation afforded to the farmers for first developing and subsequently maintaining such landraces. Satisfactory mechanisms still need to be found for ensuring that farmers benefit from genetic resources. These should be applicable not only to the initial supply of material but also to the resultant product. Above all, access should be made with prior informed consent.

Bioprospecting

The search for pharmaceuticals and other natural products has acquired real commercial dimensions. Pharmaceutical companies are rushing into developing countries to gain access to IK and biological resources in order to develop new medicines and natural products. They are eager to undertake botanical collections. There is even the presumption that this is what the countries in the south want and need. Based on the premise that many developing countries are "gene rich but technology poor," it is sometimes naively assumed that the southern countries have no option, as they need the money to alleviate their poverty or undertake conservation efforts, and that this is the only way they can obtain funds. Other terms are applied—such as "adding value to the resources"—and adding value is seen as happening only in the north. But when one considers the case of the neem, which every peasant family uses as medicine in India, one wonders what value is being added elsewhere. On the contrary, this concept merely provides property rights to the individual company abroad while depriving the poor in India of the resource, which ends up as "improved" medicine from the West that would be even more expensive to import.

Unfortunately, some unscrupulous tactics have been used. There is the very sad story of an herbalist in West Africa who had developed what appeared to be a cure for AIDS. In addition to local patients, he had cured a German national who traveled to West Africa for the treatment. Somehow, some Japanese heard about this and invited the herbalist to Japan. He must have trusted that they were going to work out an arrangement together. But, one evening, he came back to his hotel room to find it ransacked and all his information stolen. He was so disappointed that he took the first flight home, and so devastated that upon reaching home he committed suicide. Unless the Japanese got the plant or materials that the herbalist was using for treatment, this represents an opportunity completely lost to the world.

While there is much in bioprospecting to benefit humanity, there is no reason for the benefits to be one-sided. The idea of the continued availability of resources to present and future generations, often inherent in the environmental perceptions of indigenous peoples, has to be kept in mind. This embodies the principle of sustainability. When one considers that about 80 percent of the people in developing countries depend on traditional medicine, and that Western medicine is out of reach either because of distance or cost, the continued supply of medicinal plants in the wild is a more dependable alternative. If there is enough material for everybody, then contracts that would ensure equitable benefits to the local communities will need to be put into place. Moreover, there should be an option for the local people to say no to the extraction and commercialization of their resources.

The Biodiversity Treaty

The Biodiversity Treaty recognized the importance of the contributions of indigenous communities through their knowledge to biodiversity conservation. It also emphasized the value of the resources of which they are custodians. The terms by which different groups would access and make use of this legacy were left to be worked out. These remain as points for spirited discussion at national and international forums in which governments, development agencies, scientists, and community workers negotiate the terms of access to, and the distribution of benefits from, the biological resources. Against this background, what hope is there for the future, especially as related to indigenous peoples and local communities, their knowledge, and their resources?

The fact that IK is disappearing is of great concern. Some organizations are trying hard to document IK, notably the Center for Indigenous Knowledge for Agriculture and Rural Development (CIKARD) at Ames, Iowa. There is also a growing network of other centers for IK in different countries with the aim of collecting and documenting such knowledge for development. When considering the collection and documentation of IK, one encounters some underlying issues touching on the responsibilities of the collectors of information and the intellectual and resource rights of the owners. There have been efforts to work on a code of conduct for collecting information; examples can be found in Cunningham (1993) and Martin (1995). These spell out the responsibilities of ethnobiologists as they document local knowledge and practices. Such undertakings are considered in line with the recognition of indigenous intellectual property rights.

The Handbook for Indigenous, Traditional and Local Communities on Traditional Resource Rights, developed by Darrell Posey and his group, will go a long way in forming a basis for this partnership. In addition, the International Society of Ethnobiology (ISE) is trying to address some of the concerns expressed here. The society was formed at the first International Congress of Ethnobiology held in 1988 in Belem, Brazil. At the congress, a declaration known as the Declaration of Belem was passed (Posey et al. 1990). It depicts the society's aspirations in recognizing the indigenous and local communities, their knowledge and resources, and their important role in the conservation of cultural and biological diversity. It also outlines the responsibilities of ethnobiologists in seeing that the communities benefit from access to their knowledge and resources. To underline these responsibilities, the declaration is reproduced here:

Declaration of Belem

As ethnobiologists, we are alarmed that

Since

. . . tropical forests and other fragile ecosystems are disappearing, many species, both plant and animal, are threatened with extinction, indigenous cultures around the world are being disrupted and destroyed;

and Given

. . . that economic, agricultural, and health conditions of people are dependent on these resources, that native peoples have been stewards of 99 percent of the world's genetic resources, and that there is an inextricable link between cultural and biological diversity;

We, members of the International Society of Ethnobiology, strongly urge action as follows:

Henceforth

1. A substantial proportion of development aid be devoted to efforts aimed at ethnobiology inventory, conservation, and management programs.

2. Mechanisms be established by which indigenous specialists are recognized as proper authorities and are consulted on all programs affecting them, their resources, and their environments.

3. All other inalienable human rights be recognized and guaranteed, including cultural and linguistic identity.

4. Procedures be developed to compensate native peoples for the utilization of their knowledge and their biological resources.

5. Educational programs be implemented to alert the global community to the value of ethnobiological knowledge for human well-being.

6. All medical programs include the recognition of and respect for traditional healers and the incorporation of traditional health practices that enhance the health status of these populations.

7. Ethnobiologists make available the results of their research to the native people with whom they have worked, especially including dissemination in the native language.

8. Exchange of information be promoted among indigenous and peasant peoples regarding conservation, management, and sustained utilization of resources.

At subsequent congresses, which are held every two years, efforts are made to fulfill the vision that inspired the declaration.

Conclusion

In conclusion, we know that the indigenous way of looking at resources carries with it the responsibility of maintaining such resources for continued availability for the survival of present and future generations. Indigenous resource management practices have been geared toward sustainability of resource use. Outsiders who fail to appreciate this tend to be motivated more by self-interest in the extraction of both knowledge and biological resources than by genuine concern for the indigenous communities and natural resources.

At this point, the question "Am I my brother's keeper?" becomes relevant. Is everyone going to be working for his or her own narrow interests, or will we find a way of working together for a better world, a world with a future?

References

Cunningham, A. B. 1993. *Ethics, Ethnobiological Research, and Biodiversity*. Gland, Switzerland: WWF.

Martin, G. J. 1995. *Ethnobotany: A Methods Manual*. London: Chapman and Hall.

Posey, D. A., W. L. Overal, C. R. Clement, M. J. Plotkin, E. Elisabetsky, C. N. da Mota, and J. F. P. de Barros, eds. 1990. "Ethnobiology: Implications and Applications." *Proceedings of the First International Congress of Ethnobiology*, vol. 1 (Belem 1988).

16

Epilogue

Quo Vadis? The Promise of Ethnoecology

ROBERT E. RHOADES
JACK HARLAN

The conference that stimulated the papers for this volume was unusual in our meeting-plagued world. The organizer's format of bringing together a diverse group of people from strikingly different walks of life to discuss a common concern is often attempted but rarely succeeds. All too often, academics pontificate and politicians moralize while farmers and lay folk sit silently until called upon for a few earthy words. In this case, however, representatives of tribal organizations, governments, nongovernmental organizations (NGOs), private businesses, and universities mingled without awkwardness to openly and enthusiastically debate the frontiers of a newly re-emerging field called ethnoecology. Enthusiastically and without reservation, the participants shared "points of view" and "views from a point" in order to arrive at negotiated conclusions about future directions. Continuing in that spirit, the editor of this volume asked that this epilogue be written by two nonethnoecologists (an agronomist and an agricultural anthropologist) with a single question in mind: What value, if any, does ethnoecology hold for those who are not practitioners of this particular science? Therefore, this essay is our own "take"—as outsiders—on this fascinating intellectual no-man's-land lying precariously somewhere between science and folk wisdom, between the human mind and actualized behavior, and between cognitive theory and real world application.

Although diverse topics were debated by the participants, we will frame our analysis of the papers through a mutual personal accounting of how ethnoecology relates to our own interests, agrobiodiversity and plant genetic resources. While we realize that ethnoecology is potentially much broader and could focus on any theme about people's perceptions of their environment, most papers in the volume in fact center on farming or gardening peoples and their relation to biological resources, especially plants.

Three tasks will be taken up in this epilogue. First, we will make explicit our viewpoint on ethnoecology by tracing the subdiscipline alongside our personal experiences in agricultural research and genetic resources research over the past quarter-century. Second, we will examine the contemporary status of ethnoecology (its strengths and its shortcomings) as measured by the papers presented in this conference. Third, we will reflect on where we believe ethnoecology should move in order to play an active, positive role in preserving the world's biological resources and the cultures that nurture them.

Agrobiodiversity, Plant Genetic Resources, and Ethnoecology: A View from Personal Histories

Although we (Harlan and Rhoades) do not recall formally meeting prior to this conference, we share a common background that provides a historical touchstone for this epilogue and one upon which to frame the papers into a larger perspective. When Harlan was a young professor in agronomy at Oklahoma State University in the early 1960s, Rhoades was just finishing up a bachelor's degree in agriculture in the same land grant college. The campus setting was straight out of the pages of Jane Smiley's 1995 novel *Moo*. A great belief in the power of science and the ability of its handmaiden—technology—to feed a famine-stricken world pervaded the campus. We recall vividly how OSU scientists were working to eliminate the cattleman's dreaded screwworm with natural, environmentally safe methods, and how plant breeders at the college were contributing to new plant varieties that would revolutionize American agriculture.

For us, however, the real excitement lay in events linking our college to foreign lands. OSU was one of the first American universities to establish a research and development project in Africa. Professors and students returned from Ethiopia with tantalizing slides and stories of their experiences in agricultural improvement under a university-led Point-4 Program (later USAID). Direct American scientific involvement was envisioned as one of the most effective ways to help Third World agriculture "get moving," as Mosher (1966) later called the effort. Young U.S. scientists in land grant colleges like OSU were being exposed for the first time to food production issues beyond our borders, particularly the "adoption model" or the "transfer of technology and science" from the United States to Third World farmers. Given this setting, perhaps it was no accident that Rhoades soon found himself as one of John F. Kennedy's first Peace Corps volunteers in Nepal (1962–1964), promoting hybrid seeds in a new settlement valley, while Harlan had transferred to the University of Illinois, where his groundbreaking work in plant genetics came to fruition. These early career experiences shaped our thinking about local people and the relevance of their plant and natural resource knowledge.

By 1967, we were both caught up in the sweeping changes in world agriculture brought about by the Green Revolution. Rhoades, an East-West Center fellow, found himself affiliated with the International Rice Research Institute (IRRI), conducting his master's thesis research on the diffusion of IR-8 ("miracle rice") among Filipino farmers. That very same year, Harlan—now professor of agronomy at the University of Illinois—found himself chairing an international meeting in Rome, where the United Nations Food and Agriculture Organization (FAO) sponsored a conference on the erosion of plant genetic resources. This occasion marked the first in a series of meetings wherein the world's scientific community deliberated on measures to conserve genetic resources. Harlan—who did not set the agenda at this meeting, which lasted over a week—recalls that he merely pounded the gavel, told the delegates where the restrooms were and where the coffee could be found, and tried his best to slow down the fast talkers for the interpreters.

Just as rice scientists at IRRI had little understanding of the value of landraces or the impact that the release of high yielding varieties (HYVs) would have on agrobiodiversity, the participants in Harlan's meeting understood little of coming political issues such as intellectual property rights or the legal recognition of farmers' rights over their landraces. Given that the term "genetic resources" was coined by Sir Otto Frankel in that same year of 1967 (Witt 1985), it is understandable that few scientists fully grasped at that time the coming political importance of the wild relatives and landraces that traditional farmers had curated over the millennia (Witt 1985). Concepts expressed so freely and prominently in this volume—such as indigenous knowledge, intellectual property rights, farmers' rights—had not become popular currency. Ironically, on the same Philippine landscape where Rhoades studied the diffusion of miracle rice, but somewhat removed from mainstream agricultural extension, ethnographer Hal Conklin (1963) was busy at work crafting a new discipline he called "ethnoecology."

Meanwhile, back in the United States, the value of "exotic" germplasm was dramatically driven home by the 1972 National Academy of Sciences publication *Genetic Vulnerability of Major Crops*, a study stimulated by the southern corn leaf blight, which had destroyed almost 15 percent of the U.S. corn crop two years earlier (National Academy of Sciences 1972). During this same year, Harlan participated in several other international meetings, such as the one held at Beltsville, Maryland, where a proposal was developed leading to the establishment in 1974 of the International Board for Plant Genetic Resources (IBPGR, today the International Plant Genetic Resources Institute [IPGRI]), to be funded by the Consultative Group for International Agricultural Research (CGIAR). As in the earlier Rome meetings, little thought was given to the value of indigenous knowledge and traditional farming strategies underlying agrobiodiversity. Appreciation of the coevolved knowledge base had not proceeded hand in hand with the growing appreciation of the raw germplasm itself. Local people were certainly important to collecting expeditions to replenish northern gene banks or for breeding programs, but only to the degree that they could point out where materials could be collected or to help provide simple "passport" data to the materials being collected.

While ethnoecology was largely absent from the early debates on the coevolution of humans and domesticated plant biodiversity, advances in another branch of anthropology—prehistoric archaeology—was unraveling new insights into humankind's farming heritage and its role in fostering genetic diversity. The "origins of agriculture" archaeological expeditions into the Zagros-Taurus Arc in the 1950s, 60s, and 70s, and later in Mexico, provided fertile hypotheses for a better understanding of the role of Neolithic farmers in the manipulation of genetic materials over long periods. Through interaction with archaeologists and paleobotantists, Harlan and his plant scientist colleagues came to better appreciate that the very genetic materials coveted for crop improvement are also a human-cultural product crafted over the millennia (Harlan and Zohary 1966). The progenitors of modern crops had long been manipulated, transplanted between zones, cross-fertilized, and selected for certain traits, by farmers. These studies showed in the Near East that people and plants had coevolved and that in the ecotonal, mountainous environments of early domestication, the domesticated crops tended toward increasing diversity under human influence, especially as the material was carried to new environments (Rhoades 1978).

Today, the areas of greatest genetic variability of the major food crops (in terms of landraces and wild relatives) tend to be in regions of strong cultural variation. The interplay of culture and agrodiversity continued intensely through the Age of Discovery, wherein world conquest by Europeans also meant a dramatic spreading over the globe and readaptation to new settings of domesticated crops from alien lands. Not until the late nineteenth and early twentieth centuries and the spread of industrial agriculture did the value of landraces and wild species become so obvious that funded expeditions were launched to discover and conserve plant genetic resources from centers of origin and domestication.

Despite the proliferation of hypotheses on origins of agriculture, researchers could only speculate on the "ethnoecology" of the early plant experimenters. While the work of Braidwood (1960) and later Flannery (1965) provided valuable inferences regarding the mechanisms and systems (population, climate, genetic, technological) that provide a fertile ground for the intimate relationship between humans and domesticated plants, they could do little to explore the relation between perception and behavior, the twin forces crucial in the domestication of plants and animals. Harlan (1995:25) hinted at the value of ethnoecology's potential for linking behavior and cognition with knowledge when he argued that something was amiss in the "origins of agriculture quest": "One problem I have with all the published models is that they are all conceived by middle-class, university-educated, Industrial Age pragmatists, and all look for some golden bottom line that will explain it all. Input-output studies, optimum foraging strategies, and a variety of armchair theories are all products of the modern mind-set. Could we not come closer to reality if we consulted some Darwinian 'wise old savage'?"

Ethnoecology could have brought us closer to this reality if it had been engaged earlier in these cross-disciplinary research questions. Until this conference, we had separately wondered why this integrating perspective of ethnoecology has not fulfilled its promise. The cultural ecology and ecological anthropology of Julian Steward, Roy Rappaport, Robert Netting, and Kent Flannery (to cite but a few examples from the 1960s and 1970s) spoke more directly to our interests in understanding both agricultural change and issues of biodiversity. However, the papers in this symposium strongly suggest that a new synthesis is finally emerging, and this one has the potential to speak to a wider, more diverse audience than the earlier ecological anthropology and ethnoscience.

Ethnoecology's Persistence and Promise

Despite the potential for linking outward, ethnoecology's early days were highlighted by attempting to show the value of its perspective for ethnography itself. Although Conklin coined the term "ethnoecology" in the context of a 1957 FAO report on shifting cultivation—and in doing so forever changed our attitudes about "primitive farmers"—by the early 1960s, ethnoecologists were more interested in fine-tuning ethnoscience, the "methodology of new ethnography," than in using their newly discovered insights to address problems that interested other disciplines, including the agricultural sciences (Frake 1962).

Nevertheless, while the primary early audience of ethnoecologists seemed to be other ethnographers (although its founder, Conklin, never lost his cross-disciplinary base), early ethnoecological research was opening insights into what other anthropologists would come to call "indigenous knowledge," or local knowledge. Even though most of these early ethnoscientific studies did not go beyond the presentation of native categories and descriptions of relationships, they served to systematically document the complexities of taxonomic classification by then-dubbed "primitive peoples" which, in its systematicity and rigor, equaled anything in the scientific world (Berlin 1992).

Even in anthropology itself, ethnoecology was a little-known perspective. Ford (this volume) noted that the early work of Conklin (1957) and Frake (1962) argued for the incorporation of native perspectives into ecological anthropology, but this direction was largely ignored until lately by anthropologists, who were more interested in using scientific ecology (ecosystems analysis and evolutionary ecology) in their analysis. One can only speculate what the shape of anthropology would be today if the ethnoecological approach had taken the upper hand in the 1960s and 1970s. However, the very fact that ethnoecology now seems to be gaining momentum is an indication of its power in the study of human-environment interactions.

Let us now turn to the symposium papers presented in this volume—comprising the most recent as well as the largest gathering of ethnoecologists to date—and analyze how well they rise to the modern challenge. Four major themes on ethnoecology can be identified from the papers and discussions. None of these themes alone is new. However, it is the manner in which they are being combined for the first time that creates "emergent properties," as well as the promise of what lies ahead in the exciting field of ethnoecology.

First, the papers drive home the long-argued anthropological truth that, over time, many marginal, tribal peoples have created finely adapted and ecologically sound ways of living with their natural surroundings. Eugene Hunn argues poignantly in his paper that subsistence is not merely an economic activity for traditional peoples, as outsiders would believe, but a creation growing from a long-term relationship with their environment, a relationship bound up in social and religious meaning. Devon Peña shows how cultural landscapes of Chicano family farms combine with the underlying community land tenure systems to guard and enhance genetic diversity of landraces. Michael Dove not only places the complex characteristics of traditional agronomy in contemporary space but also uses ethnoecology to show how among the people he studied certain plants are ritualistically maintained as a mechanism for preserving cultural history. This finding is certainly an interesting confusion for those who see in situ conservation in traditional societies as merely a matter of market-driven, rational economics. Finally, Lillie Lane's moving essay on the practical and religious meaning of her people's hogan illustrates how cosmovision, architecture, practical needs, and social organization are woven together in the full expression of a folk ethnoecology. Second, the rich knowledge of these cultures—and the underlying cognitive patterns—are just as important for sustainable development as the more tangible material or technological aspects. Scott Atran, in his study of commons breakdown in Mayaland, points out that the roles of information and communication through social networks are overlooked in resource management efforts. By analyzing how perception of the landscape varies by ethnicity, age, and gender, Virginia Nazarea demonstrates that even in a single watershed there is a rich multihued tapestry of dramatically different points of view on resource management. Daniela Soleri and Steven Smith, in comparing plant breeders' and local populations' viewpoints on in situ conservation of landraces, point toward the differences in world views that must be resolved before significant progress can be made in both modern breeding and landrace preservation.

Third, these cognitive bases of human lifescapes and their role in shaping the natural world need to be better studied, as does the relationship between local adaptations and how they culminate to effect global environmental change. Although not directly the main topic of any single paper in the symposium, the issue of local-global connections (the role of indigenous peoples in global change) was an underlying and consistent theme. Several papers addressed issues of the commons (Hunn, Atran, Peña, Lagrotteria and Affolter), the relationship and global importance of in situ versus ex situ conservation of plant genetic resources (Dove, Soleri and Smith), the health and nutrition of indigenous peoples (Johns), and self-determination for indigenous peoples in a legalistic world (Posey, Kabuye, Stephenson, and Moran). Although ethnoecologists began in the 1950s with the modest goals of improving the method of ethnography, their findings—and institutional support as activists of indigenous peoples—have thrust their work center stage, especially in the global debate over access and rights to land, water, and genetic resources.

Fourth, scientists, policy makers, and planners alike in their drive to define and create a sustainable planet earth must listen to and join hands with those marginal, often powerless, and threatened traditional peoples who offer "different takes" on our common future (and different ways of seeing the problems and solutions). Until this symposium, we have had the distinct impression that ethnobiology and, by implication, ethnoecology (as an outgrowth of ethnoscience), have been very pedantic and plagued by anthropology's greatest sin, a cowardly disdain for application in solving real world problems. However, virtually every paper in this volume reveals that the new ethnoecologists are far from timid when it comes to declaring the value of their research, the rights of the people they work with, and the need to use local wisdom in creating a better world. While ethnoecology may have lost earlier opportunities to make a difference, the issues of today—intellectual property rights, biodiversity treaties, legal wrangling over germplasm ownership, erosion of indigenous knowledge, conflicts over traditional land and water rights—cry out for an engaged and theoretically sophisticated ethnoecology. All the evidence now indicates that a not so insignificant number of ethnoecologists are coming out and declaring the real possibility of an ethnoecology that is relevant in today's world. The pursuit of intellectual trivia for its own sake has come to an end.

Conclusion: Quo Vadis?

As outside discussants, we give the papers in this volume our highest accolades. We feel that finally ethnoecology is maturing; it is speaking outward, and with a new confidence in its global relevance. Does that mean we feel the job is done? No. It would be a mistake for ethnoecology to rest on its laurels now. The real work has only begun.

Just a few years back, we found the models of ethnolinguistic classification of plants and animals to be fascinating and titillating, but hardly useful for the practical tasks we faced (such as plant conservation and improvement and agricultural development). Ethnoecology, as expressed in the papers in this volume, has certainly lifted us above the straightjacketed approaches of static linguistic analysis to deal head-on with the social distribution and cultural contexts of knowledge, issues of time and space, and the political ecology of cognition, as well as practical application. Nevertheless, the fundamental challenge lies ahead.

Ethnoecology needs to systematically formulate a theory about how people perceive, organize knowledge about, and then act upon their environment. To provide an integrated, holistic vision has been the dream and challenge of social thinkers at least since the ancient Greeks. Humanity needs, and deserves, a theory of knowledge that ties cognition to action, information to behavior, and human action to environment. But this theory, once formulated, must be articulated for the rest of the world: for fellow scientists, for policy makers and change agents, for the lay public concerned with the future of the planet, and—finally—for indigenous peoples themselves. Once this is accomplished, ethnoecologists can set aside their tools and wait for the harvest of their accomplishments.

References

Berlin, Brent. 1992. *Ethnobiological Classification: Principles of Categorization of Plants and Animals in Traditional Societies*. Princeton: Princeton University Press.

Braidwood, R. J. 1960. "The Agricultural Revolution." *Scientific American*. 203:130–48.

Conklin, Harold. 1957. "Hanunoo Agriculture: A Report on an Integral System of Shifting Cultivation in the Philippines." Rome: FAO.

Flannery, K. V. 1965. "The Ecology of Early Food Production in Mesopotamia." *Science* 147:1247–56.

Frake, Charles. 1962. "The Ethnographic Study of Cognitive Systems." *Anthropology and Human Behavior*. Washington, D.C.: Anthropological Society of Washington.

Harlan, Jack. 1995. *The Living Fields: Our Agricultural Heritage*. New York: Cambridge.

Harlan, J., and D. Zohary. 1966. "Distribution of Wild Wheats and Barley." *Science* 153:1079.

Mosher, Arthur T. 1966. *Getting Agriculture Moving: Essentials for Development and Modernization*. New York: Praeger.

National Academy of Sciences. 1972. *Genetic Vulnerability of Major Crops*. Washington, D.C.: NAS Publications.

Rhoades, Robert E. 1978. "Archaeological Use and Abuse of Ecological Concepts and Studies: The Ecotone Example." *American Antiquity* 43(4):608–14.

Smiley, Jane. 1995. *Moo*. New York: Knopf.

Witt, Steven. 1985. *Biotechnology and Genetic Diversity*. San Francisco: California Agricultural Lands Project.

About the Contributors

JAMES M. AFFOLTER is director of research at the State Botanical Garden of Georgia and an associate professor in the Department of Horticulture at the University of Georgia. His research interests include the conservation of medicinal plant species and the domestication and commercial production of medicinal and aromatic plants. Additional long-term interests include conservation biology and management of endangered species in the southeastern United States and systematic and floristic treatments of Latin American *Apiaceae* (Parsley family).

SCOTT ATRAN is a cognitive anthropologist at the Centre National de la Recherche Scientifique, CREA-Ecole Polytechnique in Paris and a faculty associate in the Department of Psychology and the Institute for Social Research at the University of Michigan. His research focuses on how people categorize and reason about the world. A key issue of his long-term fieldwork in the Middle East and Mesoamerica is how people put to use this knowledge in maintaining or destroying their common resources and environment.

MICHAEL R. DOVE is professor of social ecology, School of Forestry and Environmental Studies, Yale University. His research interests focus on interaction between local communities, national governments, and global agencies concerning the use of natural resources. He has spent over twelve years in south and southeast Asia (principally Indonesia and Pakistan) studying these issues. Current research projects include the relative impact on biodiversity of local communities and extralocal institutions, the history of market linkages in the tropical forest, and the sociology of environmental science.

RICHARD I. FORD is professor of anthropology and biology, curator of ethnology, and director of the Ethnobotanical Laboratory, Museum of Anthropology, University of Michigan. In recognition of his academic accomplishments, he has been awarded the Arthur F. Thurnau Professorship, the Fryxell Award from the Society for American Archaeology, and the Janaki Ammal Medal (India), and elected fellow of the American Academy of Arts and Sciences. His research interests revolve around the ethnobotany of southwestern pueblos, the prehistoric cultural topology in northern New Mexico, the archaeohistoric origins of the Jicarilla Apache, and the ethnoecology of traditional southwestern societies.

JACK HARLAN, professor emeritus of agronomy at the Plant Genetics Department, University of Illinois, is an internationally renowned scholar in the areas of plant breeding and crop-human coevolution. Prior to coming to Illinois, he worked for sixteen years at Oklahoma State University, where he launched his career in plant collecting and introduction. For his life's work he was awarded the coveted Frank Meyer Medal in 1976 by the United States Department of Agriculture for meritorious contributions to plant introduction. Dr. Harlan presently lives in New Orleans.

EUGENE S. HUNN is a professor of anthropology at the University of Washington. He studied the ethnobiology and cultural ecology of the Sahaptin-speaking peoples of the mid-Columbia River basin in eastern Washington and Oregon, and his 1990 book, *Nch'i-Wana "The Big River": Mid-Columbia Indians and their Land*, prepared in collaboration with James Selam, Yakama elder, won a Washington State Governor's Writers award. He was also involved with research for the National Parks Service on rural subsistence in Alaska. Recently he has initiated an ethnobiological investigation in San Juan Mixtepec, a Zapotec-speaking village in Oaxaca, Mexico.

TIMOTHY JOHNS is associate professor of human nutrition and of plant science at McGill University and associate director of the Centre for Indigenous People's Nutrition and the Environment (CINE). He is an ethnobotanist who has worked in Latin America and Africa. His research takes a human chemical ecological approach to issues such as traditional use of plants for food and medicine in relation to health and nutrition, sustainable use of traditional plant resources, and the evolution of diet and medicine. He is the author of *With Bitter Herbs They Shall Eat It: Chemical Ecology of the Origins of Human Diet and Medicine* (1990), *Luo Biological Dictionary* (1998, with J. O. Kokwaro), and *Functionality of Food Phytochemicals* (1997, with J. T. Romeo), in addition to more than fifty scientific papers.

CHRISTINE S. KABUYE was for twenty-three years botanist-in-charge of the East African Herbarium at the National Museums of Kenya in Nairobi. She extended her work from plant taxonomy to ethnobotany and working with local communities. For four years, she also led the Kenya Resource Center for Indigenous Knowledge. Her interest in the application of indigenous knowledge to development issues was recognized within the International

Society of Ethnobiology, of which she was elected president for the years 1995–96. She has been widely involved in conservation and the utilization of biological resources at the local and international levels. She hopes to continue with the same interests during her retirement in Uganda.

MARTA LAGROTTERIA is a biologist and professor at the Universidad Libre del Ambiente in Córdoba, Argentina, where she teaches courses in the production, marketing, and conservation of medicinal and aromatic plants. She has been studying the harvest and marketing of native herbs in the province of Córdoba since 1984, and has coordinated several rural development projects sponsored by the provincial and municipal governments to encourage sustainable production of herbal products.

LILLIE LANE is from Bodaway at the westernmost end of the Navajo Reservation in Arizona. She belongs to the T'obaahi clan and is born for the Itizitani. She studied English and linguistics at the University of Arizona in Tucson. Lane worked with the Navajo Nation Historic Preservation Department in Window Rock, Arizona, before coming to the Museum of Indian Arts and Culture in Santa Fe, New Mexico. She has been serving as a consultant with help from the McCune Foundation.

KATY MORAN is the executive director of the Healing Forest Conservancy, a nonprofit foundation founded by Shaman Pharmaceuticals, Inc., to develop and implement a process to return benefits to native communities for their contribution to ethnobotanical drug discovery after a product is commercialized. Earlier, she worked for members of the U.S. House of Representatives on human rights and environmental issues. Moran was also a research collaborator at the Smithsonian's National Zoo, which supported her ethnozoological thesis research on traditional elephant management in Sri Lanka.

VIRGINIA D. NAZAREA is associate professor of ecological anthropology at the University of Georgia, where she established and directs the Ethnoecology/Biodiversity Laboratory. Previously, she was assistant professor at the Board of Environmental Studies at the University of California, Santa Cruz, and at the College of Human Ecology, University of the Philippines at Los Banos. Her research focuses on the interface between cultural and genetic diversity and on the political ecology of cognition. She has written *Local Knowledge and Agricultural Decision Making in the Philippines: Class, Gender, and Resistance* (Cornell University Press, 1995), and *Cultural Memory and Biodiversity* (University of Arizona Press, 1998).

DEVON G. PEÑA is director of the Rio Grande Bioregions Project, a research network of forty scholars, farmers, and sustainable agriculture advocates dedicated to the study of cultural and biological diversity in the Rio Grande watershed. He is the author of *The Terror of the Machine: Technology, Work, Gender, and Ecology on the U.S.-Mexico Border* (University of Texas Press, 1997) and *Chicano Culture, Ecology, Politics: Subversive Kin* (University of Arizona Press, 1998).

DARRELL A. POSEY is director of the Programme for Traditional Resource Rights of the Oxford Centre for the Environment, Ethics, and Society at Mansfield College, University of Oxford. He is also science director for the Institute for Ethnobiology of the Amazon and a graduate professor at the University of Maranhão, Brazil. Posey has been an active and respected crusader for many years in the areas of ethnobiology, ethnoecology, and indigenous rights. He is recognized as a leading proponent of traditional resource rights.

ROBERT E. RHOADES, professor of anthropology, University of Georgia, is best known for his work in agricultural and ecological anthropology. Throughout his career he has sought ways to bring the anthropological perspective to issues of food, biodiversity, and agrarian systems. He spent ten years at the International Potato Center in Lima, Peru, before moving to Asia in 1988, where he established the User's Perspective with Agricultural Research (UPWARD) Network. In 1991, he came to the University of Georgia as head of the Department of Anthropology, with a mandate to build a first-rate program focused on ecological and environmental anthropology.

STEVEN E. SMITH is an associate professor in the Department of Plant Sciences at the University of Arizona. His research experiences focus on the development of plant breeding procedures for increasing stress tolerance and the description and use of plant populations that have been directly affected by farmer selection. With Daniela Soleri he has developed and tested techniques for rapid estimation of genetic variation using data from farmers' fields. He is also investigating the possible consequences of land management practices on genetic diversity and evolutionary potential in plants native to the Sonoran Desert.

DANIELA SOLERI is a doctoral student in the Arid Lands Resource Sciences Program in the University of Arizona's College of Agriculture. She has conducted research on household gardens in drylands with David A. Cleveland. Since 1992 her research interests have focused on investigations of small-scale farmer-breeders' goals for their crop populations, the consequences of their selection and management on the genetic structure of those crop populations, and the implications for collaboration between farmers and plant breeders. She is currently pursuing those interests in Oaxaca, Mexico, with traditional farmers and their maize populations.

DAVID J. STEPHENSON, JR., practices law in Colorado. He has had extensive experience litigating computer software trade secret cases and serving as legal counsel for Native Americans in housing and intellectual property matters. He also counsels nonprofit groups based in Africa on intellectual property and human rights issues. He has both a law degree and a Ph.D. in anthropology. He has authored several articles on the intellectual property rights of indigenous peoples and on international human rights law. He has participated actively as an applied anthropologist on the Board of the Center for Cultural Dynamics and with affiliated organizations that are concerned with cultural dynamics and socioeconomic development locally, nationally, and internationally.

Index

About the Author

VIRGINIA D. NAZAREA is Professor of Anthropology at the University of Georgia and the author of *Cultural Memory and Biodiversity* (University of Arizona Press, 1998), *Local Knowledge and Agricultural Decision Making in the Philippines: Class, Gender, and Resistance*, and *Yesterday's Ways, Tomorrow's Treasures: Heirloom Plants and Memory Banking*.

CPSIA information can be obtained at www.ICGtesting.com
Printed in the USA
LVOW13s0013231213

366469LV00003B/7/P